An Introduction to Testing for Drugs of Abuse

An Introduction to Testing for Drugs of Abuse

William E. Schreiber

Clinical Director of Chemistry, LifeLabs
Professor, Department of Pathology and Laboratory Medicine
The University of British Columbia
Vancouver, Canada

WILEY Blackwell

Registered Offices
John Wiley & Sons, Inc., 111 River Street, Hoboken, NJ 07030, USA
John Wiley & Sons Ltd, The Atrium, Southern Gate, Chichester, West Sussex, PO19 8SQ, UK

Editorial Office
9600 Garsington Road, Oxford, OX4 2DQ, UK

For details of our global editorial offices, customer services, and more information about Wiley products visit us at www.wiley.com.

Wiley also publishes its books in a variety of electronic formats and by print-on-demand. Some content that appears in standard print versions of this book may not be available in other formats.

Library of Congress Cataloging-in-Publication Data
Names: Schreiber, William Edward, 1954– author.
Title: An introduction to testing for drugs of abuse / William Edward
    Schreiber, University of British Columbia, Vancouver, Canada.
Description: Hoboken, NJ : Wiley, 2022. | Includes bibliographical
    references and index.
Identifiers: LCCN 2021058984 (print) | LCCN 2021058985 (ebook) | ISBN
    9781119794059 (paperback) | ISBN 9781119794066 (adobe pdf) | ISBN
    9781119794073 (epub)
Subjects: LCSH: Drug testing.
Classification: LCC HV5823 .S45 2022 (print) | LCC HV5823 (ebook) | DDC
    615.7/040724–dc23/eng/20220107
LC record available at https://lccn.loc.gov/2021058984
LC ebook record available at https://lccn.loc.gov/2021058985

Cover Design: Wiley
Cover Image: © luchschenF/Adobe, Kondor83/Shutterstock, Stephen Orsillo/ Shutterstock, monticello/Shutterstock

Set in 9.5/12.5pt STIXTwoText by Straive, Pondicherry, India

Printed in Singapore
M115750_240222

Contents

Preface   *vii*

Section I   Substance Use, Drug Metabolism, and the Testing Process   *1*

1   **Introduction**   *3*

2   **How the Body Handles Drugs**   *13*

3   **Specimen Collection**   *19*

4   **Screening Tests: Immunoassays**   *27*

5   **Confirmation Tests: Chromatography and Mass Spectrometry**   *37*

Section II   Individual Drugs   *47*

6   **Cocaine**   *49*

7   **Amphetamines**   *59*

8   **Benzodiazepines and Z-drugs**   *77*

9   **Other Sedative-Hypnotic Drugs**   *93*

10   **Opioids**   *105*

11 **Cannabinoids** *125*

12 **Hallucinogens** *137*

13 **Alcohols** *151*

Section III   **Appendices** *177*

Appendix A   **How to Read a Toxicology Report** *179*

Appendix B   **Guideline Documents: Pain Management and Addiction Medicine** *183*

Index *185*

Preface

Toxicology is one of the most complex fields in laboratory medicine, and it is a frequent source of consultations. Doctors, nurses, patients, and other clients ask me a lot of questions about drug testing and what the results mean. Here are some examples.

- Can passive exposure to marijuana smoke cause a positive test for cannabis?
- Which drugs are detected by your amphetamine assay?
- If my patient is taking oxycodone, why was the screening test for opiates negative?
- How long can heroin be detected in a urine sample?
- Could this level of cocaine in blood be fatal?

It has taken years to learn the answers and, perhaps more importantly, to locate the resources where those answers can be found. There is a huge literature on drugs of abuse and their measurement. However, it is hard to find a publication that summarizes both the testing process and the interpretation of test results. That is what laboratory and clinical practitioners need, and that is the subject of this book.

My goal is to provide a readable account of the most common drugs of abuse – what they are, how they work, therapeutic uses, potential for abuse, and how laboratories test for them. Enough practical information is included so that readers can understand and interpret the results of drug tests. The book is an introduction, not a compendium – more detailed information is available in textbooks, journal articles, and websites, some of which are listed at the end of each chapter.

Case studies are included with all of the chapters on drugs to add interest and bring the subject to life. Some of the cases describe real people and events, while others are the product of my imagination, often based on personal experiences or stories related by colleagues. Whether real or fictional, every case poses questions to the reader, discusses the answers to those questions, and ends with several key points about drug use and the role of testing. The cases include material that is not covered elsewhere – don't skip them!

This book was conceived and written with three audiences in mind. The first group includes pathologists, clinical laboratory scientists, and medical technologists. They are the professionals who perform drug testing, report the results, and provide consultation on what those results mean. A second group consists of physicians, nurses, pharmacists, and other healthcare providers who use test results in making decisions about patient management. The third audience is composed of trainees – residents in pathology and primary care specialties as well as students in medicine, pharmacy, nursing, and other health-related programs.

I would like to acknowledge the contributions of others who have helped in the creation of this work. Dr Maria Issa and Dr Serguei Likhodi read portions of the manuscript and offered helpful feedback and suggestions. Kate Campbell produced most of the illustrations in the first section of the book, many of which describe the testing process. Finally, I salute the scientific and technical staff in the clinical laboratories of Vancouver General Hospital, the British Columbia Provincial Toxicology Centre, and LifeLabs. They have been my work colleagues for more than three decades and have taught me much of what I know about this field.

William E. Schreiber
June 23, 2021

Section I

Substance Use, Drug Metabolism, and the Testing Process

1

Introduction

This is a book about drugs that affect mood, perception, and consciousness. Drugs can relax you, energize you, take away your pain, help you to sleep, or create feelings of euphoria. When taken in excess or for long periods of time, these same drugs can injure or kill you.

What Is a Drug?

A drug is a substance that produces a change in biological function when consumed. If you look up the word "drug," you are likely to see two types of definitions:

- a substance intended for use in the diagnosis, cure, mitigation, treatment, or prevention of disease (Source: US Food and Drug Administration glossary of terms)
- an illegal substance that some people smoke, inject, etc. for the physical and mental effects it has (Source: Oxford Advanced American Dictionary).

These definitions highlight the therapeutic properties of drugs on the one hand, and their potential for inappropriate use on the other.

How Drugs Work

Small amounts of a drug can have potent effects on the body. In most cases, those effects are mediated through binding of the drug to receptors that control cellular processes.

Drugs that affect the central nervous system act on neurons to increase or decrease the transmission of nerve impulses. Some drugs affect the flow of ions

An Introduction to Testing for Drugs of Abuse, First Edition. William E. Schreiber.
© 2022 John Wiley & Sons Ltd. Published 2022 by John Wiley & Sons Ltd.

into and out of neurons, which changes their ability to initiate or propagate an action potential. Other drugs affect the release and reuptake of neurotransmitters, thereby stimulating or inhibiting communication between individual neurons.

Any drug can be used inappropriately. However, certain drugs are more prone to misuse than others because of their mood-altering or intoxicating effects. These drugs activate the brain's reward system by stimulating the release of dopamine in several pathways. The sense of euphoria they produce is accompanied by other effects – feelings of energy and power in the case of stimulants, or relaxation and peace in the case of opioids.

Drug Misuse and Abuse

The World Health Organization defines drug misuse as "Use of a substance for a purpose not consistent with legal or medical guidelines, as in the non-medical use of prescription medications." Drug or substance abuse refers to consumption of illicit drugs or to chronic use of a drug that interferes with a person's normal activities. It can involve illegal drugs purchased on the street (e.g., heroin), prescription drugs taken in excessive doses (e.g., benzodiazepines), or legal intoxicants (e.g., ethanol).

There is a blurry line between use and abuse for many drugs. When taking a substance causes harm, either to the individual or others, it can be considered substance abuse. Some of the consequences of drug abuse are:

- mental/physical disabilities
- long-term health problems
- risky behavior
- not meeting responsibilities
- altered social relationships.

A study published in *The Lancet* assessed the harm associated with 20 drugs that are commonly abused. The authors considered physical harm, dependence, and social harms to arrive at a mean score for each drug. Heroin and cocaine ranked first and second on the list of harmful drugs. In a subsequent article, the authors identified heroin, crack cocaine, and methamphetamine as the most harmful drugs to individuals and alcohol as the most harmful substance to others (Figure 1.1).

Terminology

The words *drug* and *substance* are both used in the medical community – they are generally equivalent in meaning. Substance is a broader term. It includes compounds that we may not consider to be drugs, such as alcohol, nicotine and caffeine, and it often appears in discussions of chemical dependency.

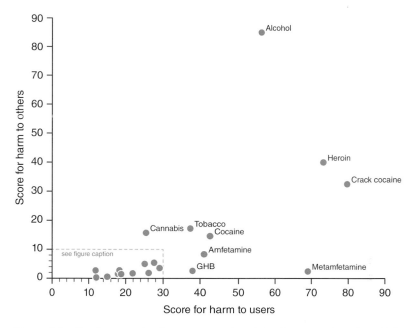

Figure 1.1 Graph comparing harm to users (x-axis) vs harm to others (y-axis) for 20 drugs and substances. Higher scores indicate a greater degree of harm. Drugs with the highest total scores are labeled. The points in the lower left corner of the graph represent (in order of decreasing harm to users) ketamine, benzodiazepines, mephedrone, methadone, butane, ecstasy, anabolic steroids, khat, LSD, mushrooms, and buprenorphine. The overall scores for each drug were based on assessments of 16 harm criteria. *Source:* Adapted from Nutt et al. (2010) with permission from Elsevier.

Increasingly, the word *abuse* is being replaced with the terms *misuse* or simply "use." Drug abuse carries a social stigma that may prevent people with chemical dependency issues from seeking help. The National Institute on Drug Abuse has produced a handout, called Words Matter, with tips on how to discuss addiction in objective, non-judgmental language. It is available at the following website: https://www.drugabuse.gov/sites/default/files/nidamed_words_matter_terms.pdf

Economic Impact of Drug Abuse

The financial impact of drug abuse on a nation's economy is staggering. According to the Surgeon General's Report on Alcohol, Drugs and Health, substance misuse and substance use disorders in the United States cost more than $400 billion per year in lost productivity, healthcare expenses, and the criminal justice system. A more recent appraisal by a provider of addiction treatment services raises that figure to $578 billion for the year 2016.

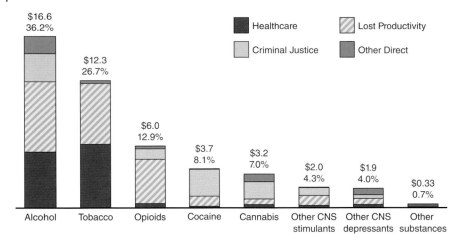

Figure 1.2 Costs of substance use in Canada in 2017. For each substance, the estimated cost is given (a) in billions of dollars and (b) as a percentage of the total cost. The costs attributable to healthcare, lost productivity, the criminal justice system, and other direct costs are indicated. *Source:* Adapted with permission from the Canadian Centre on Substance Use and Addiction, 2021.

The overall cost of substance use in Canada in 2017 was estimated at $46 billion. This amount was attributed to lost productivity (43.5%), healthcare (28.4%), criminal justice (20.1%), and other direct costs (7.9%). Figure 1.2 shows the absolute and relative cost for each substance – more granular detail is provided on pages 8–10 of the report.

Controlled Substance Schedules

Government agencies classify drugs and certain chemical compounds according to their acceptable medical use as well as their potential for abuse and harm. In the United States, the Drug Enforcement Administration (DEA) lists drugs under five distinct categories or schedules. The definition of each schedule and some examples are given in Table 1.1. A more complete list of controlled substances is available on the DEA website:

www.deadiversion.usdoj.gov/schedules/

In Canada, the Controlled Drugs and Substances Act is the national law that governs production, sale, import, export, and possession of controlled substances and precursors. Individual drugs and drug groups are listed in a series of appended schedules. The Act and its schedules appear on the Justice Laws website of the Government of Canada:

https://laws-lois.justice.gc.ca/eng/acts/C-38.8/

Table 1.1 Controlled substance schedules in the United States.

Definition	Examples
Schedule I	
Drugs with no currently accepted medical use and a high potential for abuse	Heroin
	Marijuana (cannabis)
	LSD
	peyote
	MDMA
Schedule II	
Drugs with a high potential for abuse, with use potentially leading to severe psychological or physical dependence	Cocaine
	Methamphetamine
	Methadone
	Oxycodone
	Fentanyl
Schedule III	
Drugs with a moderate to low potential for physical and psychological dependence	Codeine-containing products (90 mg or less per dosage unit)
	Ketamine
	Anabolic steroids
Schedule IV	
Drugs with a low potential for abuse and low risk of dependence	Diazepam
	Propoxyphene
	Tramadol
	Zolpidem
Schedule V	
Drugs with lower potential for abuse than Schedule IV. These preparations contain limited quantities of certain narcotics and are generally used for antidiarrheal, antitussive, and analgesic purposes	Cough preparations with codeine (200 mg or less per 100 mL)
	Lomotil® (atropine/diphenoxylate)
	Pregabalin

LSD, lysergic acid diethylamide; MDMA, 3,4-methylenedioxymethamphetamine (ecstasy).
Source: Drug scheduling, United States Drug Enforcement Administration, www.dea.gov/drug-scheduling.

Why Test for Drugs of Abuse?

There are several fields of medicine in which drug testing is an essential part of diagnosis and monitoring.

- *Emergency medicine* – patients with an acute intoxication may present with aggressive or bizarre behavior or an altered level of consciousness. Testing for alcohols and other substances in blood helps the physician to identify the cause and monitor treatment. Screening for drugs in urine can identify recent consumption of commonly abused drugs and assist in follow-up care.
- *Addiction medicine* – treatment of substance use disorders usually requires monitoring of patients for ongoing consumption of drugs. For example, patients on a methadone maintenance program are monitored to ensure that they are taking the methadone as prescribed and are not continuing to use other opioids. Treatment of alcohol use disorder includes testing for markers of alcohol consumption.
- *Pain management* – opioids and other medications are prescribed to treat chronic pain syndromes. Patients may sell or divert their medications, becoming distributors of pharmaceutical-grade drugs to others who misuse them. Regular testing is performed to ensure compliance with the treatment program and to check for use of nonprescribed medications.

Drug testing is also performed to promote safety in the workplace, support the legal system, and ensure fair competition in sports.

- *Workplace drug testing* – drug use among employees is responsible for time lost from work and increased numbers of job-related accidents. Intoxicated employees are more likely to cause injury or death to themselves, their coworkers and innocent third parties. The transportation industry regularly tests drivers, pilots, and other workers for drug use to maintain standards of safety and accountability. Cut-off concentrations for initial and confirmatory tests are published by the United States Department of Transportation (Table 1.2).
- *Forensic toxicology* – this term refers to drug analysis for legal purposes. Testing of drivers for ethanol and other drugs is used to establish intoxication, which may lead to arrest, prosecution, and civil or criminal penalties. People may be given drugs without their knowledge to incapacitate them for purposes of robbery or sex (drug-facilitated assault). Analysis of the victim's blood or urine can identify such drugs. Toxicology is also a standard part of death investigations conducted by coroners and medical examiners. Postmortem analysis for drugs in blood, urine, and vitreous specimens is routinely performed to investigate a possible drug-related cause of death.
- *Performance enhancement* – elite athletes, both professional and amateur, want a competitive edge, and some of them find it in performance-enhancing drugs.

Table 1.2 Cut-off concentrations for drugs and metabolites in urine.

Analyte	Initial test	Confirmatory test
Marijuana metabolite (THC-COOH)	50	15
Cocaine metabolite (benzoylecgonine)	150	100
Codeine/morphine	2000	2000
Hydrocodone/hydromorphone	300	100
Oxycodone/oxymorphone	100	100
6-Acetylmorphine	10	10
Phencyclidine	25	25
Amphetamine/methamphetamine	500	250
MDMA/MDA	500	250

These cut-offs specify the minimum concentration of a drug or metabolite that constitutes a positive test. All concentrations are in ng/mL. Samples with a positive initial test must undergo a confirmatory test before reporting the result. Grouped analytes such as codeine/morphine are both detected in the initial test, but they are measured and reported separately in the confirmatory test. MDA, 3,4-methylenedioxyamphetamine; MDMA, 3,4-methylenedioxymethamphetamine; THC-COOH, delta-9-tetrahydrocannabinol-9-carboxylic acid.
Source: United States Department of Transportation www.transportation.gov/odapc/part40/40-87

These drugs are taken to improve strength and stamina and are different from recreational drugs of abuse. Examples include anabolic steroids, human growth hormone, and erythropoietin. Performance-enhancing drugs are banned from major sporting competitions. The ban is enforced by testing athletes on a sporadic basis and immediately following competition.

What You Will Find in This Book

This book was written for students and practitioners who want a concise description of drugs of abuse and how clinical laboratories test for them. The emphasis is on interpretation of test results rather than analytical methods. The first section covers basic concepts of drug metabolism, specimen collection, and techniques for detecting and measuring drugs. The following section contains chapters on individual drugs or drug groups, all organized under similar headings. Practical information about laboratory tests (i.e., screening cut-offs, length of time a drug can be detected, interferences) is included. Lists of metabolites that indicate use of a particular drug are provided in Appendix A.

Resources appear at the end of each chapter – they include journal articles, book chapters, government agency reports, websites, and videos. Wherever possible, website addresses are provided to give the reader quick access.

A number of case studies are attached to each chapter on individual drugs and drug groups. They are meant to pose real-world situations in which the facts and principles discussed in this book can be applied. Cases are drawn from most of the areas mentioned above – emergency medicine, addiction medicine, pain management, workplace testing, and forensics.

General references

Drugs of abuse testing sits at the crossroads of analytical chemistry, pharmacology, toxicology, and several branches of medical practice. When searching for information about a particular drug or topic, you may need to consult a variety of sources. The following general references are recommended as a starting point.

Textbooks

Katzung, B.G., Masters, S.B., and Trevor, A.J. (eds.) (2012). *Basic and Clinical Pharmacology*, 12e. New York: McGraw-Hill.
   A classic in the field. Chapters are well organized, well written, and well illustrated. This text is worth reading to learn the basics or fill in knowledge gaps.

Baselt, R.C. (2011). *Disposition of Toxic Drugs and Chemicals in Man*, 9e. Seal Beach, CA: Biomedical Publications.
   This book presents curated information on more than 1000 drugs and chemicals. Each entry contains sections on occurrence and usage, blood concentrations, metabolism and excretion, toxicity, and analysis. It is an essential reference for toxicology laboratories.

Hoffman, R.S., Howland, M.A., Lewin, N.A. et al. (eds.) (2015). *Goldfrank's Toxicologic Emergencies*, 10e. New York: McGraw-Hill.
   Another classic, Goldfrank's is the reigning bible of toxicology in the emergency setting. Historical background, pharmacology, pathophysiology, clinical presentation, diagnosis, and management are covered for most drugs/toxins/poisons. Chapters are somewhat variable in structure and content.

Review article

Schulz, M., Schmoldt, A., Andresen-Streichert, H., and Iwersen-Bergmann, S. (2020). Revisited: Therapeutic and toxic blood concentrations of more than 1100 drugs and other xenobiotics. *Crit. Care* 24: 195.

The authors have compiled a list of therapeutic, toxic, and fatal blood concentrations for 1100+ drugs – the list of annotations and references is even longer. These tables are especially useful when reviewing postmortem toxicology results, as they can assist in determining the cause of death.

Websites

National Institute on Drug Abuse (NIDA)
  www.drugabuse.gov
    The National Institute on Drug Abuse is one of the institutes of the National Institutes of Health. Its mission is ". . . to advance science on the causes and consequences of drug use and addiction and to apply that knowledge to improve individual and public health." The NIDA website contains extensive information on drugs of abuse written in easily understood language.

Substance Abuse and Mental Health Services Administration (SAMHSA)
  www.samhsa.gov
    The Substance Abuse and Mental Health Services Administration is an agency within the US Department of Health and Human Services. Its mission is ". . . to reduce the impact of substance abuse and mental illness on America's communities." The SAMHSA website offers numerous publications, brochures, and fact sheets related to substance abuse.

Centre for Addiction and Mental Health (CAMH)
  www.camh.ca
    The Centre for Addiction and Mental Health is ". . . Canada's largest mental health teaching hospital and one of the world's leading research centers in its field." The Health Info tab on the CAMH website has educational material on drug use, addiction, and mental health topics.

Further Reading

Articles and reports

Canadian Substance Use Costs and Harms Scientific Working Group. (2020). Canadian substance use costs and harms 2015-2017. (Prepared by the Canadian Institute for Substance Use Research and the Canadian Centre on Substance Use and Addiction.) Ottawa: Canadian Centre on Substance Use and Addiction. www.ccsa.ca/canadian-substance-use-costs-and-harms-2015-2017-report.

Nutt, D., King, L.A., Saulsbury, W., and Blakemore, C. (2007). Development of a rational scale to assess the harm of drugs of potential misuse. *Lancet*. 369: 1047–1053.

Nutt, D.J., King, L.A., and Phillips, L.D. (2010). Drug harms in the UK: a multicriteria decision analysis. *Lancet.* 376: 1558–1565.

US Department of Health and Human Services (HHS), Office of the Surgeon General (2016). *Facing Addiction in America: The Surgeon General's Report on Alcohol, Drugs, and Health.* Washington, DC: HHS https://addiction.surgeongeneral.gov/ sites/default/files/surgeon-generals-report.pdf.

Websites

National Institute on Drug Abuse
Drugs, brains and behavior: the science of addiction
www.drugabuse.gov/publications/drugs-brains-behavior-science-addiction/ introduction

Videos

2-Minute Neuroscience – Reward system. www.neuroscientificallychallenged.com/ blog/2-minute-neuroscience-reward-system.

AACC Pearls of Laboratory Medicine
Drugs of abuse testing
www.aacc.org/science-and-research/clinical-chemistry-trainee-council/trainee- council-in-english/pearls-of-laboratory-medicine/2011/drugs-of-abuse-testing.

2

How the Body Handles Drugs

There are four steps in a drug's journey through the body: (i) absorption, (ii) distribution, (iii) metabolism, and (iv) excretion. These processes determine the time frame during which a drug can be detected in blood, urine, and other body fluids. They also determine the rate at which new compounds (i.e., metabolites) appear after a drug is taken. The study of drug absorption, distribution, metabolism, and excretion – what the body does to a drug – is called *pharmacokinetics*.

Absorption

Absorption is the process by which drug molecules enter the bloodstream. The rate of this process depends on the route of administration.

Oral	When taken orally, a drug is absorbed across the mucosal lining of the stomach or small intestine before entering the blood. Onset of pharmacological effects typically occurs in 15–60 minutes, depending on the properties of the drug and its formulation (i.e., tablet vs liquid).
	Following absorption, orally administered drugs are carried by the portal circulation directly to the liver, where they may be converted to less active metabolites before entering the systemic circulation. This "first-pass" effect can limit the bioavailability of a drug.
	Example: benzodiazepines
Intravenous	Intravenous (IV) injection bypasses the absorption step, as the drug is introduced directly into the circulation. The IV route provides the fastest onset of drug action.
	Example: heroin

(Continued)

An Introduction to Testing for Drugs of Abuse, First Edition. William E. Schreiber.
© 2022 John Wiley & Sons Ltd. Published 2022 by John Wiley & Sons Ltd.

Inhalation	Air-borne drug molecules are pulled into the lungs, cross alveolar walls (where gas exchange takes place) and diffuse into adjacent capillaries. This happens when a drug is smoked or inhaled as an aerosol. Onset of drug effects can be very rapid.
	Example: cannabis
Insufflation	Better known as "snorting," insufflation causes drugs to contact nasal passages, which are lined by a mucous membrane. It is a more porous barrier than skin and allows rapid diffusion of drugs into underlying blood vessels.
	Example: cocaine
Transdermal	Diffusion across skin is a slow process. Drug patches are used to deliver a constant amount of drug per unit of time and maintain stable levels in blood.
	Example: fentanyl

Other routes exist (e.g., sublingual, intramuscular or subcutaneous injection), but they are not typically used with illicit or diverted prescription drugs.

Distribution

Once a drug has entered the circulation, it is transported through the body and diffuses into tissues. This process is influenced by several factors.

- *Blood flow* – drugs are more rapidly distributed to areas of the body with a high rate of blood flow. For most drugs of abuse as well as many prescription medications, the main target organ is the brain, which is well perfused (15% of cardiac output in the resting state).
- *Chemical structure* – drugs that are lipophilic and are not ionized more readily cross lipid bilayers and enter cells.
- *Protein binding* – when bound to plasma proteins (mainly albumin), drugs cannot freely pass through capillary walls and enter tissues. The degree of plasma protein binding therefore affects the amount of drug that reaches its target.

Drugs that are very lipid soluble may distribute into adipose tissue, accumulate there, and eventually diffuse back into the circulation. This can prolong the action of the drug, and it increases the length of time a drug or its metabolite(s) can be detected in blood or urine.

Metabolism

The purpose of metabolism is to transform lipid-soluble compounds, which readily enter tissues, into a more water-soluble form that can be excreted. This occurs in two phases.

- *Phase I* – drug molecules are chemically altered by oxidation, reduction, or hydrolysis. These reactions are catalyzed by enzymes in the liver and, to a lesser extent, in other tissues. The most important drug-metabolizing enzyme system, cytochrome P450 (CYP), contains multiple isoforms that act on a wide variety of drugs. The resulting products are usually inactive or less active than the parent drug. However, in some cases the metabolite has similar or even greater potency.
- *Phase II* – metabolites from phase I, or in some cases the unchanged drug, are conjugated to a water-soluble group. This usually involves the addition of glucuronic acid or sulfate to an available –OH on the drug molecule (Figure 2.1).

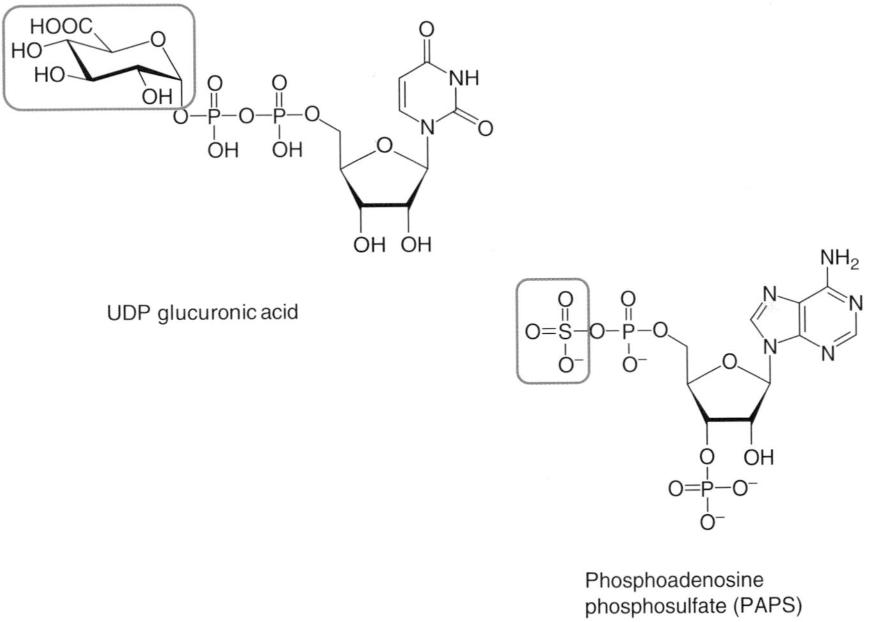

UDP glucuronic acid

Phosphoadenosine phosphosulfate (PAPS)

Figure 2.1 Structures of uridine diphosphate (UDP) glucuronic acid and phosphoadenosine phosphosulfate (PAPS), which serve as donors for glucuronic acid and sulfate in phase II reactions. The glucuronic acid and sulfate groups are indicated by rectangles. Addition of either group to a drug molecule increases its water solubility and facilitates excretion in urine and/or bile.

The reactions are catalyzed by UDP glucuronosyltransferase and sulfotransferase enzymes, respectively.

The rate of metabolism varies among individuals. Genetic polymorphisms in CYP and other drug-metabolizing enzymes affect their activity and can accelerate or slow down the rate of drug transformation. People who are slow metabolizers may experience toxicity from a drug at doses considered therapeutic. Conversely, rapid metabolizers require more drug to reach therapeutic levels in blood.

Excretion

Drugs and their metabolites are primarily excreted by two routes: (i) glomerular filtration into urine and (ii) transport into bile.

The kidneys remove most drugs and their metabolites from the body. The pH of the glomerular filtrate affects the excretion of weakly acidic and basic drugs, because ionized molecules are not reabsorbed by renal tubules. Molecules that are smaller and more water soluble are usually excreted in this way.

Biliary excretion requires active transport of drugs and metabolites out of liver cells and into the biliary system. Bile flows into the duodenum, and its contents are ultimately discharged in feces. Drugs that are larger (molecular weight [MW] > 300) and more lipophilic are preferentially excreted in bile.

Other routes of excretion exist but are less important. Volatile compounds can diffuse out of capillaries in the alveolar wall and enter the air spaces of the lungs, from which they are exhaled (e.g., ethanol). Excretion of drugs in breast milk may affect infants who are breastfeeding.

Disease of the kidneys or liver can impair excretion, causing drug levels in blood and other body fluids to rise.

Further Reading

Book chapter

Correia, M.A. (2012). Drug biotransformation. In: *Basic and Clinical Pharmacology*, 12e (eds. B.G. Katzung, S.B. Masters and A.J. Trevor), 53–68. New York: McGraw-Hill.

Websites

Pharmacology Education Project
www.pharmacologyeducation.org/clinical-pharmacology/clinical-pharmacokinetics

MSD Manual

Drug Absorption
www.msdmanuals.com/professional/clinical-pharmacology/pharmacokinetics/
drug-absorption

Drug Bioavailability
www.msdmanuals.com/professional/clinical-pharmacology/pharmacokinetics/
drug-bioavailability

Drug Distribution to Tissues
www.msdmanuals.com/professional/clinical-pharmacology/pharmacokinetics/
drug-distribution-to-tissues

Drug Metabolism
www.msdmanuals.com/professional/clinical-pharmacology/pharmacokinetics/
drug-metabolism

Drug Excretion
www.msdmanuals.com/professional/clinical-pharmacology/pharmacokinetics/
drug-excretion

Videos

Pharmacokinetics 1 – Introduction
www.youtube.com/watch?v=8-Qtd6RhfVA

Pharmacokinetics 2 – Absorption
www.youtube.com/watch?v=pWW-aq7iSa0

Pharmacokinetics 3 – Distribution
www.youtube.com/watch?v=6erefsWCVxg

Pharmacokinetics 4 – Metabolism
www.youtube.com/watch?v=ztsBn8gsfHw

Pharmacokinetics 5 – Excretion
www.youtube.com/watch?v=VZRVt9r4oSM

3

Specimen Collection

Testing for drugs and toxins can be performed on a variety of specimens. Choosing the most appropriate one depends on the purpose of the analysis.

Types of Specimens

Blood

Blood is the specimen of choice for assessing the amount of active drug in a patient. This information is required in a number of situations, such as:

- monitoring the concentration of a prescribed drug to ensure therapeutic levels are present (eg, *carbamazepine in a patient with epilepsy*)
- measuring the amount of a potentially toxic drug to guide therapy and assess prognosis (eg, *acetaminophen in a patient who took an overdose*)
- identifying a cause for altered mental status (eg, *ethanol in a comatose patient*).

In most cases, drugs are measured in serum. Whole blood is required for certain analyses, because the drug or toxin of interest accumulates within red blood cells.

Collecting blood is an invasive procedure – inserting a needle through skin into a vein – and it causes anxiety and discomfort to some patients. Experienced phlebotomists can draw blood with little or no associated pain, and they ensure that the sample is from the person on whom it was ordered.

An Introduction to Testing for Drugs of Abuse, First Edition. William E. Schreiber.
© 2022 John Wiley & Sons Ltd. Published 2022 by John Wiley & Sons Ltd.

Urine

Urine is the best sample for determining whether someone has taken a drug within the recent past. It is an aqueous fluid that contains both drugs and their metabolites in a concentrated form. The window of detection varies from hours to a week or longer, depending on the specific drug and how much was taken.

Urine drug testing is ordered for a number of reasons.

- Confirm or rule out a suspected acute drug intoxication (eg, *patient in the emergency room with symptoms of a drug ingestion*).
- Discover use of illegal or prohibited drugs (eg, *employment drug screening program*).
- Monitor patients with a history of drug abuse (eg, *drug rehabilitation program*).
- Ensure compliance with drug prescriptions that might be diverted (eg, *opioid screening in pain management clinics*).

Urine is easy to collect and does not require special handling. Point-of-care screening tests can be performed directly on the sample with no additional preparation. Most compounds of interest are excreted in urine, so recent consumption of a drug is likely to be detected in this sample.

The main drawback of a urine sample is that it can be tampered with by the owner before analysis. The issue of specimen validity testing (SVT) is discussed later in this chapter.

Oral Fluid

Oral fluid consists mainly of the mixed saliva that comes from three major and several minor salivary glands. Small amounts of cellular debris, bacteria, and the residue of ingested substances are also present. It is more than 99% water and contains electrolytes, proteins, enzymes, and other biomolecules.

Drugs appear in oral fluid by passive diffusion from blood. The rate of transfer is related to the chemical properties of a drug and the degree of binding to plasma proteins (only free drug molecules can diffuse across capillary walls).

The window of detection is several hours to several days following drug use, depending on the specific drug and the amount and frequency of consumption. One advantage of oral fluid over other samples is that recently ingested (e.g., ethanol) or smoked (e.g., tetrahydrocannabinol) drugs may leave a residue that is present in high concentration.

Oral fluid is a noninvasive specimen and is easy to obtain with commercial collection devices. The collection process can be witnessed to prevent sample adulteration or substitution. It is especially useful in roadside testing for drugs that impair a person's ability to drive.

Hair

Drugs in blood diffuse into the hair follicle, where the hair shaft grows. As the keratin matrix of hair is formed, drugs are incorporated and remain there, moving outward as the length of the hair shaft increases. A human hair is therefore a biological record of drug exposure.

The unique advantage of hair testing is the extended time period over which drugs can be detected. Hair samples can test positive as early as 1 week following drug use, and the window of detection is months or even years, depending on hair length.

Hair is usually collected from the crown of the head. The average growth rate for hair at this site is about 0.5 in (1.3 cm) per month. Sample collection is fairly simple – multiple strands of hair are cut at scalp level and sent to a testing facility for analysis. Collection can be witnessed or performed by trained personnel to prevent substitution or adulteration of the hair sample.

The laboratory may test a defined length of hair, which corresponds to an approximate period of time (i.e., hair trimmed to 1.5 in will detect drugs taken within the past 3 months). It is also possible to cut hair into short segments, which are analyzed separately. By comparing the test results on adjoining hair segments, an approximate timeline of drug exposure can be determined.

Analysis of drugs in hair is technically more demanding than other specimens. Hair samples must be washed to eliminate any surface contamination, and drugs must then be extracted from the hair shaft into a liquid medium that can be analyzed. This is a specialized procedure that is not available in most routine toxicology laboratories.

Other Specimens

Other sample types are informative in specific situations.

- *Meconium* is the greenish-black stool passed by a newborn following delivery. Analysis of this material can detect drug exposure in utero (i.e., drug use by the mother during pregnancy).
- *Sweat* can be collected with an absorbent skin patch for up to 1 week. Analysis of the accumulated sweat will indicate whether drugs have been consumed during that time period.

Specimen Validity Testing

Many people are required to undergo drug testing as a condition of employment or to remain in drug rehabilitation programs. Testing is most often performed on urine because it is easy to obtain, and both drugs and their metabolites are present in concentrated form.

How Cheaters Try to Beat Drug Tests

Collection of urine samples is not usually witnessed. The person providing the sample may try to get a "clean" test result by tampering with the specimen before submitting it. Specimen tampering can be divided into three categories.

- *Adulteration* – a substance is added to the specimen that masks the presence of drugs and their metabolites. The substance may interfere with the analytical method or it may react with the drug(s) of interest to produce compounds that are not detected. Bleach, vinegar, ammonia, and other household products have been used for this purpose. More sophisticated operators use additives that interfere with drug tests but are hard to detect. These adulterants include nitrite, chromate, peroxide, halogens (all of which are oxidizing agents), and glutaraldehyde. Commercial products with amusing names like Whizzies, UrineLuck, and Stealth are available on the internet and in drug paraphernalia shops.
- *Substitution* – the patient substitutes another sample for their own urine. It could be a friend's urine, a synthetic urine product, a yellow-colored fluid such as apple juice, or tap water. Since the sample does not come from the patient, it will test negative (unless they pick the wrong person to provide the substitute sample!).
- *Dilution* – the goal of this strategy is to dilute drugs in urine below the threshold for a positive result. This can be accomplished in several ways: (i) drinking an excess of fluids prior to collection, (ii) taking diuretics to flush drugs out of the urinary tract, and (iii) adding water or another liquid to the sample itself.

Assessing Validity

Assessment of specimen validity begins with measuring urine temperature, which should be 90–100 °F (32–38 °C) within 4 minutes of collection. The physical appearance is also evaluated for color and consistency, and any unusual odor is noted.

The following chemical tests are run on patient samples to detect tampering that may give a false result.

Creatinine	Creatinine is excreted in urine and serves as a marker for this fluid. Random urine specimens contain at least 20 mg/dL (1.8 mmol/L) of creatinine. Values below this level may indicate a dilute or substituted sample
Specific gravity	This test measures the number of solute particles in urine. The reference range for specific gravity is 1.002–1.030. A low specific gravity may indicate a dilute or substituted sample
pH	Normal human urine has a pH between 4.5 and 9.0. Adulteration of a sample with an acid or base will alter the pH

Oxidant	Tests for oxidants can detect chromate, nitrite, bleach, peroxide, and iodine. Normal human urine does not contain oxidants above the assay cut-off, and increased values may indicate adulteration
Nitrite	Nitrite is not detectable in normal urine samples, but it may be elevated in patients with a urinary tract infection. High levels of nitrite may indicate an adulterated sample
Chromate	Chromate is an effective oxidizing agent, and it is not normally present in human urine. A positive test suggests that the urine sample has been adulterated
Aldehyde	Glutaraldehyde and other aldehydes are not present in normal human urine. A positive test suggests that the sample has been adulterated

Interpretation of Specimen Validity Testing

The Substance Abuse and Mental Health Services Administration (SAMHSA) has developed criteria for specimen validity in workplace drug testing programs. Dilute and substituted specimens are identified by comparing the values for creatinine and specific gravity (Table 3.1). To label a specimen as adulterated, an abnormal pH or the presence of a known adulterant is required (Table 3.2).

Table 3.1   Substance Abuse and Mental Health Services Administration criteria for labeling a urine specimen as dilute or substituted.[a]

Test	Dilute	Substituted
Creatinine	≥ 2 and <20 mg/dL	<2 mg/dL
Specific gravity	>1.001 and <1.003	≤ 1.001 OR ≥ 1.020

[a] Both criteria (creatinine and specific gravity) must be met.

Table 3.2   Substance Abuse and Mental Health Services Administration criteria for labeling a urine specimen as adulterated.[a]

Test	Adulterated
pH	<3.0
	≥ 11.0
Nitrite	≥ 500 µg/mL
Chromium (VI)	Present
Glutaraldehyde	Present

[a] Any one criterion is adequate. Depending on how samples are tested, other adulterants may be detected as well.

Table 3.3 Substance Abuse and Mental Health Services Administration criteria for labeling a urine specimen as invalid.[a]

pH value $\geq$3.0 and <4.5 and $\geq$9.0 and <11.0
Nitrite value $\geq$200 and <500 µg/mL
Inconsistency between creatinine and specific gravity
Interference in a screening or confirmatory assay
Presence of oxidizing compounds
Possible presence of:
• chromium (VI)
• halogen (e.g., iodide)
• surfactant (e.g., soap)
Physical appearance of specimen

[a] Individual laboratories may use different or additional criteria.

Specimens can also be deemed invalid in a number of situations (Table 3.3). An invalid result may indicate specimen tampering, but other explanations are possible. A review of the patient's history and consultation with the laboratory may clarify the situation.

Further Reading

Articles

Allen, K.R. (2011). Screening for drugs of abuse: which matrix, oral fluid or urine? *Ann. Clin. Biochem.* 48: 531–541.

Bosker, W.M. and Huestis, M.A. (2009). Oral fluid testing for drugs of abuse. *Clin. Chem.* 55: 1910–1931.

Cooper, G.A.A., Kronstrand, R., and Kintz, P. (2012). Society of hair testing guidelines for drug testing in hair. *Forensic Sci. Int.* 218: 20–24.

Dasgupta, A. (2007). The effects of adulterants and selected ingested compounds on drugs-of-abuse testing in urine. *Am. J. Clin. Pathol.* 128: 491–503.

Verstraete, A.G. (2004). Detection times of drugs of abuse in blood, urine and oral fluid. *Ther. Drug Monit.* 26: 200–205.

Book chapter

Garg, U. and Cooley, C. (2019). Testing of drugs of abuse in oral fluid, sweat, hair, and nail: analytical, interpretative, and specimen adulteration issues. In: *Critical Issues in Alcohol and Drugs of Abuse Testing*, 2nd Edition (ed. A. Dasgupta), 405–427. Academic Press.

Technical assistance publication

Clinical Drug Testing in Primary Care (2012). HHS Publication No. (SMA) 12-4668. Rockville, MD: Substance Abuse and Mental Health Services Administration. https://store.samhsa.gov/sites/default/files/d7/priv/sma12-4668.pdf.

4

Screening Tests: Immunoassays

Most screening tests for illicit and prescription drugs are performed by immuno-assays. The concentrations of many therapeutic drugs, as well as common over-the-counter analgesics, are also measured with immunoassays.

The popularity of this technique is due to several factors.

- *Specificity* – assays detect only a single drug or class of related drugs.
- *Sensitivity* – drugs can be measured at low concentrations.
- *Flexibility* – immunoassays can be run on automated analyzers in a central labo-ratory, producing large numbers of results with minimal effort and expense. They can also be formulated as a single test on a manual device for point-of-care applications.
- *Technical simplicity* – in either automated or point-of-care testing, running an immunoassay does not require special expertise.
- *Speed* – individual samples can be analyzed in less than an hour.

How Immunoassays Work

Immunoassays are based on the ability of antibodies to recognize and bind to a target antigen in a biological specimen. There are two formats: competitive and noncompetitive.

In a *competitive* assay, limited amounts of antibody and a labeled antigen are mixed with the patient sample. The unlabeled antigen in the sample competes with the labeled antigen for binding sites on the antibody. As the amount of anti-gen (i.e., drug) in the sample increases, less of the labeled antigen binds to the antibody, and vice versa. The amount of antigen in the sample will be either directly or inversely proportional to the signal produced by the label.

An Introduction to Testing for Drugs of Abuse, First Edition. William E. Schreiber.
© 2022 John Wiley & Sons Ltd. Published 2022 by John Wiley & Sons Ltd.

In a *noncompetitive* assay, two antibodies that recognize different epitopes on the antigen are mixed with the patient sample. One antibody is attached to a surface and captures the antigen. The other antibody, which contains a label, binds to a different part of the antigen. The amount of signal produced by the label is directly proportional to the amount of antigen in the patient sample. This format is often referred to as a sandwich assay.

Drug molecules are too small to measure with sandwich assays, because a drug–antibody complex interferes with binding of the second antibody. Immunoassays for drugs use the competitive format.

Labels for Competitive Immunoassays

The first immunoassays used radioactivity as the label. *Radioimmunoassays (RIA)* are based on competition between antigen labeled with a radioisotope (usually iodine-125) and native antigen in the sample to be tested. After all components have been added and allowed to incubate, antibody-bound label is separated from unbound label by precipitation, and its radioactivity is measured. The proportion of bound radioactivity is calculated and compared to a calibration curve to determine the amount of antigen in the sample. This type of assay, which requires separation of bound from unbound label, is called a *heterogeneous* assay (Figure 4.1).

In the 1980s, *enzyme immunoassays* were developed in which an enzyme serves as the label. When the labeled antigen–enzyme complex is free in solution, it retains enzymatic activity and produces a signal. Upon binding to an antibody, the enzyme's catalytic activity is reduced, and it produces little or no signal. Following incubation of antibody, labeled antigen, and patient sample, substrate is added and enzyme activity is measured. There is no need to separate bound from unbound label, so this type of assay is called a *homogeneous* assay.

Enzyme and other nonisotopic immunoassays for drugs have replaced RIAs in clinical laboratories. RIAs have a limited shelf-life, because radioisotopes decay over time, and handling and disposing of radioactive material require special precautions. They also require separation of bound and unbound label, an extra step that is unnecessary with homogeneous assays.

Types of Assays

Many clever variants of competitive, homogeneous drug assays have been developed. The assays described below are the ones most commonly used in clinical laboratories.

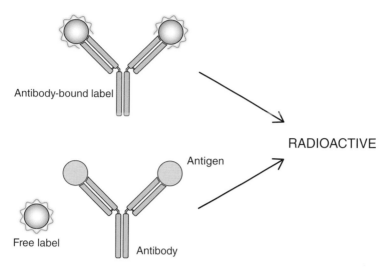

Antibody-bound label

RADIOACTIVE

Antigen

Free label

Antibody

Figure 4.1 Radioimmunoassay (RIA). A drug labeled with a radioisotope competes with unlabeled drug (antigen) in the patient sample for antibody binding sites. Both the antibody-bound and free labels are radioactive, so bound and free label must be separated before measuring radioactivity in counts per minute. RIA is an example of a heterogeneous assay.

Enzyme Multiplied Immunoassay Technique (EMIT)

The EMIT assay is based on competition for antibody-binding sites between drug contained in the sample and drug labeled with an enzyme. When the drug–enzyme complex binds to the antibody, enzymatic activity is reduced. Therefore, the more drug in a sample that binds to the antibody, the more drug–enzyme complex that remains free and the greater the enzymatic activity (Figure 4.2).

Enzyme multiplied immunoassay technique assays use glucose-6-phosphate dehydrogenase (G6PD) as the enzyme label. G6PD converts nicotinamide adenine dinucleotide (NAD) to the reduced form (NADH), which absorbs light at 340 nm. By measuring the increase in absorbance at this wavelength, the amount of drug in the sample can be determined.

Cloned Enzyme Donor Immunoassay (CEDIA)

Cloned enzyme donor immunoassay is based on the enzyme beta-galactosidase, which has been genetically engineered into two inactive fragments, one smaller (donor) and one larger (acceptor). These fragments reassociate spontaneously in solution to form a fully active enzyme. The drug of interest is covalently attached to the donor fragment. When bound to antibody, the donor can no longer associate with the acceptor fragment, and there is no enzyme activity (Figure 4.3).

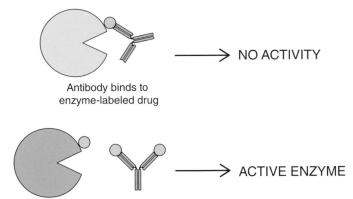

Antibody binds to
enzyme-labeled drug

NO ACTIVITY

ACTIVE ENZYME

Free enzyme-labeled drug

Figure 4.2 Enzyme multiplied immunoassay technique (EMIT). An enzyme attached to the drug serves as the label. When free in solution, the drug–enzyme complex has full activity; when bound to antibody, the activity is greatly reduced. There is no need to separate free from bound enzyme when measuring the amount of drug in a patient sample.

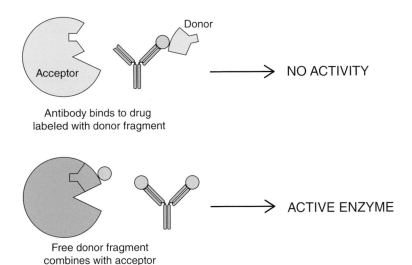

Donor

Acceptor

Antibody binds to drug
labeled with donor fragment

NO ACTIVITY

ACTIVE ENZYME

Free donor fragment
combines with acceptor

Figure 4.3 Cloned enzyme donor immunoassay (CEDIA). The free donor–drug complex combines with the acceptor fragment to form an active enzyme. Binding of antibody to the donor–drug complex prevents association with the acceptor fragment, and no enzyme activity is generated.

If drug is present in the sample, it occupies antibody-binding sites and allows the two fragments of beta-galactosidase to associate into an active enzyme. The change in absorbance reflects beta-galactosidase activity and is proportional to the concentration of drug in the sample.

Kinetic Interaction of Microparticles in Solution (KIMS)

In the KIMS assay, the drug of interest is covalently attached to microparticles. Antibodies to the drug cross-link the microparticles into larger aggregates, which increase the turbidity of the solution. Drug molecules in the patient sample occupy antibody-binding sites and inhibit cross-linking of microparticles (Figure 4.4).

The degree of microparticle aggregation, and thus turbidity, is monitored by spectrophotometry. The amount of turbidity is inversely proportional to drug concentration in the patient sample.

Why Immunoassays Sometimes Fail

Immunoassays are usually designed to detect one specific drug, metabolite, or class of drugs. Most of the time the answers are reliable, but occasionally they are not.

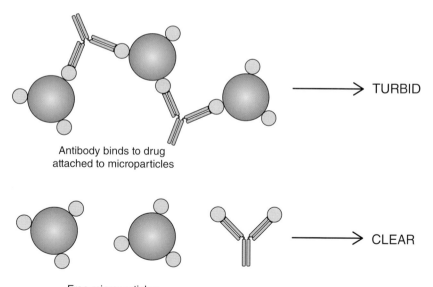

TURBID

Antibody binds to drug
attached to microparticles

CLEAR

Free microparticles

Figure 4.4   Kinetic interaction of microparticles in solution (KIMS). The drug to be measured is attached to the surface of microparticles. Antibodies to the drug cross-link microparticles into larger aggregates, increasing the turbidity of the solution. As the amount of drug in a patient sample increases, fewer microparticles are cross-linked and turbidity decreases.

Cross-reactivity

Antibodies in the test kit may cross-react with compounds other than the drug being measured. This may be due to structural similarities within a class of drugs (e.g., amphetamines) or recognition of an epitope in an unrelated drug. Assay manufacturers list drugs that do and do not cross-react in a sheet that accompanies each kit. These lists are long but not exhaustive, so recent publications and well-versed colleagues should be consulted when an unexpected result is encountered. *Cross-reactivity is a significant cause of false-positive results.*

Antibody Interference

Some patient specimens have antibodies that interfere with the assay system. *Heterophilic antibodies*, which occur naturally, can bind to antibodies in the test kit and cause inaccurate results. *Human antianimal antibodies* may develop in people who work or live with animals or who have been treated with therapeutic antibodies raised in an animal. Human antimouse antibodies (HAMA) are the most common type. *Autoantibodies* are found in patients with autoimmune disorders and may interfere with some assays. Since antibodies are plasma proteins, this type of interference is not a problem when testing urine samples.

Matrix Effect

The term *matrix* refers to the components of a sample other than the analyte being measured. Samples may contain high levels of an enzyme or metabolite that affects the performance of enzyme immunoassays. Drugs or supplements taken by the patient may interfere with assay reagents and spectrophotometric detection systems.

Prozone

Also called the hook effect, prozone refers to a large excess of antigen that saturates the assay antibody and causes a falsely low or negative result. This occurs in one-step sandwich immunoassays that are designed to measure larger antigens (e.g., proteins), so it does not usually affect drug assays.

When testing for drugs of abuse, a positive immunoassay is presumptive evidence of a drug's presence. Confirmation by a second method may not be required for clinical management, but it is compulsory when a test result is used for legal purposes.

Point-of-Care Immunoassays

Drug testing can also be performed on single-use devices that are read visually by the operator. Point-of-care testing (POCT) allows healthcare providers to collect samples, analyze them, and make decisions about patient management during a

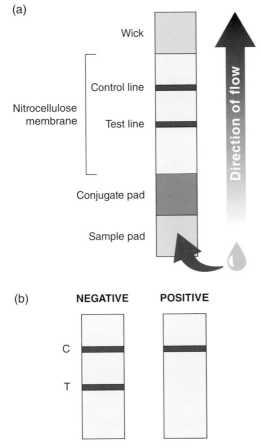

Figure 4.5 Lateral flow immunoassay. (a) A patient sample is applied to the sample pad and flows through the conjugate pad toward the test and control lines. The conjugate pad contains antibodies attached to a colored label, usually gold nanoparticles or latex microparticles. Drug molecules are immobilized on the test line of the nitrocellulose membrane. If drug is present in the sample, it occupies binding sites on the labeled antibody and prevents it from reacting with drug molecules on the membrane, so no line is formed. (b) Appearance of the test (T) and control (C) lines when the drug test is negative (left) and positive (right). A visible control line indicates that the sample has moved across the strip and the embedded reagents are working properly.

single visit. The testing device consists of a strip, cassette or cup – no additional equipment is required. Following analysis, the testing device is discarded.

Point-of-care testing for drugs is performed by lateral flow immunoassays. A liquid sample is applied to one end of the test strip and flows through it by capillary action. As it does so, fluid in the sample picks up labeled reagents embedded in the strip and carries them past immobilized reagents on the test (T) and control (C) lines. Binding of labeled reagents to the immobilized material produces a colored line. The absence or presence of a colored test line determines whether the result is positive or negative. A colored line in the control area indicates that the test is working properly (Figure 4.5).

Point-of-care testing devices can analyze a single drug or they can incorporate multiple drug assays into one cassette or cup. The assays are designed to give either a positive or negative result, based on a predetermined cut-off value for each drug. Those cut-offs are often based on guidelines from the Substance Abuse and Mental Health Services Administration (SAMHSA). Most POCT for drugs is performed on urine samples, but other fluids (e.g., oral fluid, serum) may be tested in this way.

There are numerous kits on the market for detecting drugs by POCT. They may differ in the cut-off concentration for a positive test, the analyte measured (parent drug vs metabolite), and the degree of cross-reactivity with other drugs. When using POCT kits for drugs, it is important to review these characteristics of the assay before interpreting test results.

Further Reading

Articles

Koczula, K.M. and Gallotta, A. (2016). Lateral flow assays. *Essays in Biochem.* 60: 111–120.

Moeller, K.E., Kissack, J.C., Atayee, R.S. et al. (2017). Clinical interpretation of urine drug tests: what clinicians need to know about urine drug screens. *Mayo. Clin. Proc.* 92: 774–796.

Reisfield, G.M., Salazar, E., and Bertholf, R.L. (2007). Rational use and interpretation of urine drug testing in chronic opioid therapy. *Ann. Clin. Lab. Sci.* 37: 301–314.

Saitman, A., Park, H.-D., and Fitzgerald, R.L. (2014). False-positive interferences of common urine drug screen immunoassays: a review. *J. Anal. Toxicol.* 38: 387–396.

Book chapter

Datta, P. (2019). Immunoassay design for screening of drugs of abuse. In: *Critical Issues in Alcohol and Drugs of Abuse Testing*, 2nd Edition (ed. A. Dasgupta), 121–127. Academic Press.

Videos

How does a lateral flow immunoassay work?
www.abingdonhealth.com/videos/how-does-a-lateral-flow-immunoassay-work.

AACC Pearls of Laboratory Medicine

Methodology: Immunoassays
www.aacc.org/science-and-research/clinical-chemistry-trainee-council/trainee-council-in-english/pearls-of-laboratory-medicine/2011/methodology-immunoassays.

5

Confirmation Tests: Chromatography and Mass Spectrometry

Confirmation tests for drugs are based on chromatography and, in most cases, mass spectrometry. These techniques have a number of advantages over immunoassays.

- Chromatography and mass spectrometry are versatile techniques – practically any drug or metabolite can be measured. Immunoassays are usually purchased from commercial vendors, and their test kits are limited to the more commonly encountered drugs.
- Many different drugs can be analyzed in a single assay. For example, a screening immunoassay detects only the presence or absence of opiates as a group. A liquid chromatography-mass spectrometry (LC-MS) assay can detect and measure individual opioids and their metabolites.
- Drugs can be detected at lower concentrations.
- Identification of drugs is based on their chemical and physical properties, not simply on their reaction with an antibody. This improves the reliability of assays, with few false-positive results.

Chromatography and mass spectrometry also have disadvantages when compared to immunoassays.

- Equipment is expensive. A typical cost to buy and install a new LC-MS instrument is in the range of $250–500 thousand dollars.
- Specially trained technologists are required to set up and maintain the instruments and run the assays.
- It takes hours to prepare and analyze samples and review the results. Test reports may not be available until the following day or later.

These assays are best suited to applications where an immediate answer is not required and the results must have a high degree of certainty.

The combination of chromatography with mass spectrometry has become the standard of practice for toxicology laboratories. This chapter reviews both techniques and explains how they work.

An Introduction to Testing for Drugs of Abuse, First Edition. William E. Schreiber.
© 2022 John Wiley & Sons Ltd. Published 2022 by John Wiley & Sons Ltd.

What is Chromatography?

Chromatography is a technique for separating a mixture of compounds, so that they can be identified and measured. The term *chromatography* is derived from two Greek words: *chroma* (color) and *graphein* (to write). It was coined by Mikhail Tsvet, a Russian-Italian botanist who developed the technique to study plant pigments. Although most chromatography does not involve colored compounds, the name has persisted.

Every chromatographic system has a *mobile phase* that flows past a *stationary phase*. When a chemical mixture is added to the system, molecules that do not interact with the stationary phase will be carried along with the mobile phase. Molecules that have an affinity for the stationary phase will be retained and move more slowly. The farther the mobile phase travels, the greater the separation between components in the mixture.

The chemical basis for separating drugs by chromatography is a phenomenon called *adsorption*. This refers to the attraction of molecules to a surface based on intermolecular forces, such as hydrogen bonding and dipole–dipole interactions. As the mobile phase sweeps sample components forward, molecules that interact with chemical groups on the stationary phase will be selectively retained on its surface. The strength of the interaction determines how rapidly or slowly a molecule moves along the stationary phase.

Thin-Layer Chromatography (TLC)

The principles of chromatography are illustrated by considering a specific technique. In thin-layer chromatography, the stationary phase consists of an adsorbent layer (e.g., silica gel) attached to a glass slide or plastic support. The sample is applied as a spot near the bottom edge and allowed to dry. The lower end of the slide is placed in a chamber containing a solvent (mobile phase), which moves up the adsorbent layer by capillary action. When the solvent front has traveled most of the way up the slide, it is removed from the chamber.

Sample components are visualized by treating the slide with a chemical stain or by viewing it under ultraviolet light. They appear as a series of spots between the application point and the solvent front. The distance each spot has moved relative to the solvent front gives its identity, and the relative intensity of the spot is related to its concentration in the sample. Multiple samples can be run side by side for comparison.

Thin-layer chromatography was a popular technique for detecting drugs, but it has largely disappeared with advances in immunoassay and column chromatography methods.

Gas Chromatography (GC)

In gas chromatography, the mobile phase is an inert gas, usually nitrogen or helium. The gas carries sample components through a column that contains the stationary phase. The length of time from sample injection until a component reaches the detector is called the *retention time*.

A diagram of a gas chromatograph appears in Figure 5.1. The carrier gas (mobile phase) flows from a gas cylinder into the column. At the head of the column, an injection port heats the sample to vaporize it, then transfers the vaporized sample into the column's path. During transit through the column, sample components move at different rates until they exit the column. The effluent then moves past a detector, which generates an electrical signal when sample molecules are present. Data from the detector are collected by a computer, and the amount of signal (y-axis) is graphically displayed as a function of time (x-axis) – ideally, it looks like a series of sharp peaks rising from a flat baseline. A chromatogram of blood alcohols is shown in Figure 5.2.

To develop an assay, the drugs and metabolites of interest are injected into the GC instrument and their retention times are recorded. Conditions can be varied to achieve an optimal separation of all compounds. Standards of known concentration are run through the system to calibrate the detector's response to each compound being measured. When a sample is injected, its components are identified from their retention times, and the concentrations are determined by comparison of the peak area or peak height to a standard curve.

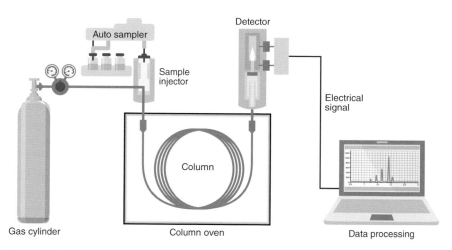

Figure 5.1 Diagram of a gas chromatograph. The carrier gas flows from the gas cylinder to the sample injector, through the column and into the detector. Individual samples can be injected manually with a syringe, or they can be placed in a tray and injected at intervals by an auto-sampler as shown here. Any of several detectors may be used to detect sample components.

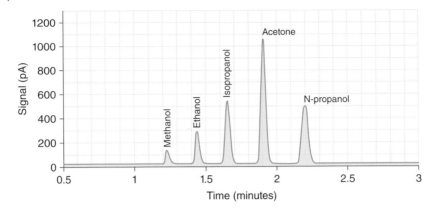

Figure 5.2 Blood alcohols separated by gas chromatography. The x-axis gives the retention time in minutes, and the y-axis measures the signal from the detector in picoamperes (pA). The last peak, n-propanol, is added to the sample as an internal standard. Acetone is a metabolite of isopropanol. *Source:* Redrawn from a chromatogram provided by Dr. Dailin Li.

A computer system controls the operation of the GC instrument and collects and analyzes data. Results for each sample, including the concentrations of all identified compounds, can be displayed on the computer screen or printed for review by others.

Columns

Capillary columns used in toxicology laboratories are long and thin – a typical length is 30 m with an inner diameter of 0.1–0.5 mm. Columns are made of fused silica, a flexible material that can be wound into coils for ease of handling. The stationary phase is either coated onto the inner surface of the column or covalently bonded to the inner surface. It can be polar or nonpolar, depending upon the type of column that is used.

The temperature of the column affects the interaction between sample components and the stationary phase, and therefore the elution pattern. To provide a temperature-controlled environment, the coiled column is mounted inside an oven. The temperature of the oven can be raised during a chromatographic run to improve the quality of separation and reduce the time required to elute all compounds.

Detectors

A number of detectors can be attached to a gas chromatograph. The two most common detectors in clinical laboratories are the *flame ionization detector* (FID) and the *mass spectrometer* (MS). In an FID, sample components leaving the column pass through a flame and produce ions. The ions generate an electrical signal that is proportional to the amount of the compound in the sample. Nearly all drugs and metabolites of interest can be detected by an FID. Mass spectrometers, which are described in a later section, provide information about the molecular mass and structure of eluting compounds.

More selective detectors are used for specific applications. A *nitrogen-phosphorus detector* (NPD) is especially sensitive to molecules containing nitrogen or phosphorus atoms. An *electron capture detector* (ECD) can detect molecules with electronegative groups, such as halogens (fluorine, chlorine, bromine, and iodine atoms).

Derivatizing Samples

One limitation of gas chromatography is that it can only analyze sample components that are volatile and heat stable. If sample components do not enter the gas phase, or if they decompose at operating temperatures, they will not be detected. For many drugs, these problems can be overcome by chemically derivatizing the sample prior to analysis. The drug is then identified in a sample from the retention time and properties (e.g., mass spectrum) of its derivative.

Liquid Chromatography (LC)

In liquid chromatography, the mobile phase is a liquid that flows through a column containing the stationary phase. The composition of the mobile phase affects the rate at which sample components elute from the column, and it can be varied during a run to optimize separations. This is different from GC, in which the mobile phase acts only as a carrier and does not modify the elution pattern. Retention times are defined in the same way for both GC and LC.

The mobile phase is pumped from a reservoir into the column, past a detector and then to a waste container. An injector system introduces sample into the mobile phase before it enters the column. Sample components move at different rates through the column, based on their relative affinity for the stationary phase. As individual components elute, they pass through a detector, which generates an electrical signal. These data are collected and displayed as peaks at various retention times (Figure 5.3).

Liquid chromatography systems in clinical laboratories use pumps to control the rate of solvent flow. This reduces the time required to perform each analysis and improves the consistency of results. The term *high-performance liquid chromatography* (HPLC) is commonly used to describe the technique. Assay development is similar to the process for GC. Standards for each compound of interest are analyzed to establish retention times and calibration curves.

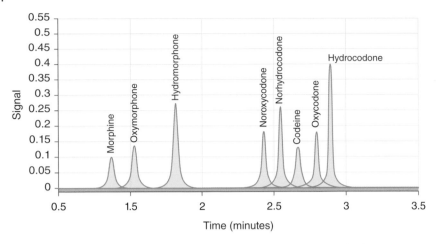

Figure 5.3 Prescription opioids and their metabolites separated by liquid chromatography-mass spectrometry. Compounds are identified by their retention times (x-axis) and mass spectral characteristics.

Stationary and Mobile Phases

The stationary phase consists of tiny beads of uniform size (typically 5 μm in diameter) that are packed into a metal column 1–5 mm wide and 5–30 cm in length. The beads are coated with a nonpolar chemical group such as C_{18}, C_8, or phenyl that has a high affinity for nonpolar compounds within a sample. Drugs and their metabolites will stick to the beads when injected as an aqueous solution. Addition of a less polar solvent will solubilize the drugs and cause them to elute. The use of a nonpolar stationary phase is called *reversed-phase chromatography*.

The mobile phase is usually a mixture of an aqueous buffer and an organic solvent, such as methanol or acetonitrile. The aqueous buffer determines the pH of the mobile phase, which affects the ionization of sample components and their affinity for the nonpolar stationary phase. The organic solvent in the mixture affects the partitioning of solutes between the mobile and stationary phases. As the proportion of organic solvent increases, sample components are more easily eluted from the column. Many HPLC instruments are equipped with two pumps that draw from two distinct reservoirs, producing a gradient of increasing organic solvent in the mobile phase as the run progresses.

Detectors

The most popular LC detectors in clinical toxicology laboratories are mass spectrometers (discussed in the next section) and absorbance detectors. *Absorbance detectors* generate a signal when molecules passing through the flow cell absorb ultraviolet (UV) or visible light. These detectors may function at a single wavelength or multiple wavelengths, depending upon the instrument design. Most drugs absorb light in the UV-visible spectrum, but many other compounds do as well. Absorbance is best suited to measuring drugs that are present in relatively high concentrations or that absorb light at a wavelength where there is little interference from other compounds.

Other detectors may be used in LC systems. A *fluorescence detector* can identify and measure compounds with fluorescent properties. An *electrochemical detector* generates a signal when a compound passing through the flow cell is oxidized or reduced. These types of detectors have high sensitivity and are useful for measuring selected drugs.

Advantages of LC

One advantage of LC over gas chromatography is that samples are analyzed in a liquid environment, so sample components do not need to be volatile. A second advantage is that both the mobile and stationary phases can be adjusted to produce an optimal separation. Nearly any drug can be measured by LC.

Mass Spectrometry (MS)

Mass spectrometry is a technique for determining the mass of molecules. When used as a detector for GC or LC, it provides mass spectral information about compounds as they elute from the column. The combination of retention time (from chromatography) and structural data (from MS) allows for highly accurate identification of drugs and their metabolites.

There are three main components to a mass spectrometer: (i) an ion source, which ionizes molecules in the sample, (ii) a mass analyzer, which separates charged molecules based on their mass-to-charge (m/z) ratio, and (iii) an ion detector, which registers a signal that is proportional to the number of ions detected. These components are depicted in Figure 5.4.

In order to be analyzed by a mass spectrometer, molecules have to be charged, because measurements are based on the movement of ions in an electric field. The mass analyzer actually determines the m/z ratio of each ion rather than its mass. Fortunately, the charge (z) on measured ions is usually 1, so the m/z values are equivalent to molecular weights.

During the ionization process, molecules collide and release fragments, which are then measured along with the intact molecule (*parent ion*). The m/z values of these fragments are reproducible and characteristic for each compound, and their signals may be stronger than that of the parent ion. The *mass spectrum* of a compound displays the m/z ratio of all ions (x-axis) and the relative abundance of each ion (y-axis) (Figure 5.5).

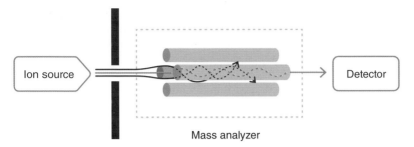

Figure 5.4 Components of a mass spectrometer. The ion source generates charged molecules and molecular fragments to be analyzed. The mass analyzer determines the mass-to-charge (m/z) ratio of each ionized species. The ion detector amplifies the signal so that the relative abundance of each ion can be measured. Data are recorded by a computer and displayed as a mass spectrum that is unique to each compound. The mass analyzer in this diagram is a quadrupole (see text for details).

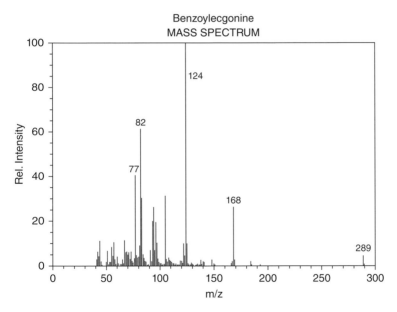

Figure 5.5 Mass spectrum of benzoylecgonine, the major metabolite of cocaine. The mass-to-charge (m/z) ratios are plotted on the x-axis and the relative abundance of each ion is plotted on the y-axis. The parent ion of benzoylecgonine has a m/z value of 289. However, the most abundant ion appears at a m/z value of 124. *Source:* Adapted from the National Institute of Standards and Technology, NIST Chemistry WebBook, https://webbook.nist.gov/cgi/cbook.cgi?ID=C519095&Units=SI&Mask=200#Mass-Spec.

A data management system collects mass spectral data from molecules that are analyzed by MS. The pattern of m/z peaks and their relative abundance is compared to a library of mass spectra for known compounds. If the system finds a match, the eluting compound is identified, and its concentration is determined from the strength of the signal.

Mass Analyzers

A number of mass analyzers, each based on different principles, have been developed. The most commonly used mass analyzers in clinical laboratories are the quadrupole and triple quadrupole.

A *quadrupole* mass analyzer is a set of four parallel metal rods to which voltages are applied, generating an electric field (see Figure 5.4). Ions produced by the ion source enter the quadrupole, where their path is affected by the electric field. If an ion passes through the quadrupole without colliding with one of the rods, it will strike a detector and produce a signal. The electric field is controlled so that only ions with a single m/z ratio can hit the detector at one time. The quadrupole functions as a mass filter.

The electric field can be set to allow ions at one or several m/z ratios to be measured, or it can be continuously varied through a range of m/z values to give a full scan of all ions. The former approach, called selected ion monitoring (SIM), is especially useful for the targeted analysis of specific drugs. Operating in the full scan mode will detect more ions, but at the cost of lower sensitivity.

A *triple quadrupole* analyzer consists of three quadrupoles (Q1, Q2, and Q3) in a linear arrangement. The Q1 quadrupole allows ions with a particular m/z ratio (*precursor ion*) to pass into Q2. This second quadrupole functions as a collision cell – the precursor ion collides with gas molecules and is fragmented into *product ions* which enter Q3. The third quadrupole is another mass filter that allows only product ions with selected m/z ratios to pass through and strike the detector. This type of instrument is also called a *tandem mass spectrometer* (MS/MS).

Analysis by MS/MS is very reliable, because a compound needs to generate both a specific precursor ion and product ion to be detected. The ion pair is called a *transition*. When only one transition is monitored, the process is called selected reaction monitoring (SRM). Two or more transitions between precursor and product ions are usually monitored to improve identification. This is known as multiple reaction monitoring (MRM).

High-resolution mass analyzers can measure m/z ratios with much greater accuracy than quadrupole-based analyzers. These instruments are better suited to untargeted analyses, where it is important to identify all possible drugs/metabolites of interest. Two instruments that produce highly accurate mass values are *time-of-flight* (TOF) and *Orbitrap*^{TM} mass analyzers (their operating principles are not covered here).

Most GC-MS instruments are equipped with a quadrupole mass analyzer. Most LC-MS instruments employ a triple quadrupole mass analyzer to detect and measure drugs. High-resolution mass analyzers are found mainly in large reference laboratories that perform drug testing.

Mass spectrometry is a complex field, and many details have been omitted in writing this general introduction to the technique. Readers can find more information about MS in book chapters, journal articles, and websites on the subject.

Further Reading

Book chapters

Hage, D.S. (2018). Chromatography. In: *Tietz Textbook of Clinical Chemistry and Molecular Diagnostics*, 6th Ed (eds. N. Rifai, A.R. Horvath, C.T. Wittwer), 266–294. Elsevier: St Louis.

Rockwood, A.L., Kushnir, M.M., Clarke, N.J. (2018). Mass spectrometry. In: *Tietz Textbook of Clinical Chemistry and Molecular Diagnostics*, 6th Ed (eds. N. Rifai, A.R. Horvath, C.T. Wittwer), 295–323. Elsevier: St Louis.

Websites

Scripps Center for Metabolomics - Basics of mass spectrometry
https://masspec.scripps.edu/learn/ms/index.html

Agilent Teaching Tools
https://www.agilent.com/en/promotions/academia-teaching

Fundamentals of gas chromatography, Agilent Technologies, 2002.
https://www.agilent.com/cs/library/usermanuals/public/G1176-90000_034327.pdf

Videos

Chromatography, animation
https://www.youtube.com/watch?v=0m8bWKHmRMM
Gas chromatography, animation
https://www.youtube.com/watch?v=iX25exzwKhI
Thin layer chromatography (TLC), animation
https://www.youtube.com/watch?v=CmHFVxTxkGs
Fundamentals of GC, Introduction and Overview (Agilent)
https://www.youtube.com/watch?v=M8d1u7kFZe0

AACC Pearls of Laboratory Medicine

Liquid chromatography: LC basics and separation techniques
https://www.aacc.org/science-and-research/clinical-chemistry-trainee-council/
trainee-council-in-english/pearls-of-laboratory-medicine/2016/liquid-chromatography-
lc-basics-and-separation-techniques
Methodology: Mass spectrometry
https://www.aacc.org/science-and-research/clinical-chemistry-trainee-council/
trainee-council-in-english/pearls-of-laboratory-medicine/2011/methodology-
mass-spectrometry

Section II

Individual Drugs

6

Cocaine

Cocaine is a naturally occurring alkaloid produced by the coca plant, *Erythroxylon coca* (Figure 6.1). This plant is cultivated in the northern Andes mountains, which provide an optimal climate for its growth. Colombia, Bolivia, and Peru produce most of the world's cocaine.

The leaves of the coca plant are treated to yield a coca paste. Cocaine is then extracted from the paste by organic solvents and dried to give the final product. There are two forms of the drug: cocaine hydrochloride (a white powder) and crack cocaine (an off-white rock or crystal).

The structures of cocaine and its major metabolites appear in Figure 6.2. All of these compounds are derivatives of ecgonine, which has a distinctive 7-carbon ring with bridging nitrogen atom. Benzoyl and methyl groups are attached to ecgonine through ester linkages.

How Cocaine Works

Within the central nervous system, cocaine blocks the reuptake of the neurotransmitters dopamine, serotonin, and norepinephrine.

Dopamine is released by specific neurons to transmit signals across the synaptic cleft. Its action is terminated when a dopamine transporter pumps dopamine back into the presynaptic neuron. Cocaine binds to the dopamine transporter and prevents the reuptake of dopamine, which leads to ongoing stimulation of the postsynaptic nerve cell. An increase in dopamine is considered to be responsible for the rewarding and addictive effects of cocaine.

Similarly, cocaine inhibits the reuptake of norepinephrine and serotonin. An increase in norepinephrine activates the sympathetic nervous system, causing a rapid heart rate, elevated blood pressure, and increased activity.

An Introduction to Testing for Drugs of Abuse, First Edition. William E. Schreiber.
© 2022 John Wiley & Sons Ltd. Published 2022 by John Wiley & Sons Ltd.

Figure 6.1 Dried leaves from the coca plant, *Erythroxylon coca*. *Source:* Ildi/Adobe Stock Photos.

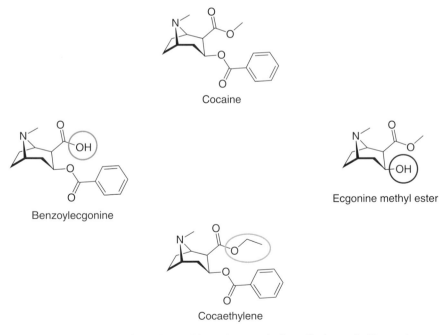

Cocaine

Benzoylecgonine

Ecgonine methyl ester

Cocaethylene

Figure 6.2 Structures of cocaine and its major metabolites. Circles and ellipses show the differences between cocaine and each metabolite.

In the peripheral nervous system, cocaine binds to voltage-gated sodium channels and inhibits sodium transport across cell membranes. This interferes with the initiation and conduction of nerve impulses and is responsible for the anesthetic properties of the drug.

Physiological Effects

The major effects of cocaine on the body are:

- excitement
- euphoria
- hyperactivity, agitation
- tachycardia
- hypertension
- increased body temperature, sweating
- dilated pupils.

Complications of cocaine use include seizures, myocardial infarction, stroke, and sudden death.

Excited delirium is a state in which the cocaine user develops severe hyperthermia, engages in bizarre or violent behavior, and demonstrates extreme strength and insensitivity to pain. Police are often called to confront these patients. Excited delirium is a medical emergency and carries a high mortality rate.

Therapeutic Uses

In medical practice, cocaine is applied topically to the interior surfaces of the nose, mouth, and throat to provide anesthesia during procedures. Cocaine is also a vasoconstrictor, which limits bleeding and helps to shrink mucous membranes. It is not available as an over-the-counter or prescription medication for individual use.

Potential for Abuse

Cocaine is typically self-administered by nasal insufflation ("snorting"), smoking or intravenous injection. When taken intravenously or smoked as crack cocaine, the drug effects begin in less than a minute, peak within 3–5 minutes, and last for up to an hour. The time course is slower when the drug is snorted, with a peak effect at 20–30 minutes and total duration up to 2 hours.

Cocaine is highly addictive for some users. The intense feelings of euphoria and energy ("rush") can lead to psychological dependence on the drug. When the effects of the drug wear off, users may experience fatigue, anxiety, depression, and mood swings ("crash"). It does not produce physical dependence, and there is no physical withdrawal syndrome as seen with discontinuation of opioids or alcohol.

In 2018, an estimated 5.5 million people in the United States aged 12 years and older (2.0% of this population) reported cocaine use in the past year. Cocaine use was higher among males (2.6%) than females (1.5%), and people aged 26–29 years were the heaviest users (6.0%). In 2017, a total of 13 942 drug overdose deaths in the United States (20% of all overdose deaths) involved cocaine.

Metabolism

The half-life of cocaine is 0.7–1.5 hours. It is metabolized to benzoylecgonine and ecgonine methyl ester, the two main metabolites, by liver enzymes (see Figure 6.2). Both compounds are pharmacologically inactive. Benzoylecgonine is also produced by the spontaneous chemical hydrolysis of cocaine. Several other metabolites are generated in smaller amounts.

In the presence of ethanol, the methyl ester group of cocaine is replaced by an ethyl group to form cocaethylene (see Figure 6.2). Cocaethylene has similar potency to cocaine and a somewhat longer half-life, which contribute to the biological and toxic effects of the drug.

Testing for Cocaine

Screening

Screening tests for cocaine are performed with immunoassays that detect the principal metabolite benzoylecgonine. For workplace drug testing (SAMHSA), the cut-off for a positive result is 150 ng/mL. A higher cut-off of 300 ng/mL is used when testing clinical samples.

Benzoylecgonine has a longer half-life than cocaine or ecgonine methyl ester, which makes it the preferred compound to monitor. Immunoassays for benzoylecgonine are highly specific. Table 6.1 gives the cross-reactivity of one commercial assay with cocaine and ecgonine.

Confirmation

For clinical testing, confirmation of a positive result is not usually required. Forensic drug testing requires analysis of cocaine and/or its principal metabolites. Both gas chromatography-mass spectrometry (GC-MS) and liquid

Table 6.1 Cross-reactivity of structurally related compounds in a commercial immunoassay for benzoylecgonine.

Compound	% Cross-reactivity
Benzoylecgonine	100%
Cocaine	<1%
Ecgonine	2.5–7.5%

Source: Siemens ADVIA Chemistry Systems, Cocaine Metabolite 2.

chromatography-tandem mass spectrometry (LC-MS/MS) methods are commonly used. These techniques can detect cocaine, benzoylecgonine, and cocaethylene in urine and other biological specimens.

Window of Detection

Screening tests may be positive for up to 3 days following cocaine ingestion when using a cut-off of 300 ng/mL. A more sensitive confirmation assay with a lower cut-off can detect cocaine use for an additional several days.

Interferences

The screening immunoassay for cocaine is very specific. One review paper (Saitman et al. J Anal. Toxicol. 2014;38:387–396) did not list any compounds that give false-positive results.

Case Studies

Case 6.1   Throat Lozenges

KS, a 37-year-old plumber, was convicted for possession of 10 g of cocaine. The judge gave him a suspended sentence on the condition that he enroll in a drug rehabilitation program. For the past year, he has attended biweekly counseling sessions and provides urine samples for drug testing once a month at an associated clinic. During his most recent clinic visit, he was seen by Dr Innocente, who had just graduated from medical school and was in her first year of residency training.

Dr Innocente reviewed the test results with the patient.

Immunoassays

Amphetamines	Negative
Cannabis	Negative
Cocaine	Positive
Opiates	Negative

Confirmation by GC-MS

Benzoylecgonine	1600 ng/mL

She said, "According to this report, you had a positive test for cocaine in your urine, but all they found on the confirmation was benzoylecgonine."

KS responded, "Oh yeah, they get this result all the time. That benzo stuff is in the throat lozenges I like to suck on." He produced a package from his right pocket and handed it to her. "See, it says right there in the ingredients – benzocaine. That's why the test came back positive."

- What is benzocaine?
- Does benzocaine interfere with screening or confirmation tests for cocaine?
- Did KS take cocaine?

Discussion

Benzocaine is a topical anesthetic that is available in many over-the-counter medications. It is the active ingredient in gels that are applied to the gums to soothe pain, aerosols that are sprayed on sunburned skin, and lozenges held in the mouth to relieve a sore throat.

Benzocaine has a different chemical structure from benzoylecgonine, the major urinary metabolite of cocaine. It does not interfere with screening tests for cocaine (which are designed to detect benzoylecgonine), nor does it interfere with GC-MS or LC-MS assays for these two compounds. However, the word *benzocaine* contains the first half of "**benzo**ylecgonine" and the last half of "co**caine**." Healthcare professionals who are not well informed might assume that benzocaine is similar to cocaine and could give a positive test result. They might even believe that benzoylecgonine is a by-product of benzocaine use.

Screening tests for cocaine are among the most reliable drug immunoassays, and the presence of benzoylecgonine by GC-MS analysis confirms the result. KS is using cocaine again.

Takeaway messages

- The names of drugs and their metabolites may be similar to the names of other medications. Don't mix them up.

- Cocaine has a short half-life. It is converted rapidly to benzoylecgonine and other products and may not be detected by confirmation assays. The presence of benzoylecgonine alone is sufficient evidence of cocaine consumption.
- Patients may volunteer their own interpretation of test reports. Use your own judgment, because they may be attempting to mislead you.

Case 6.2   The Morning After

The morning after an all-night wrap party/sleepover at a film producer's house, one of the guests could not be roused. Paramedics arrived 30 minutes later and found the assistant director on the living room floor, motionless. They checked for vital signs, began life support measures, and attached electrocardiogram (ECG) pads to monitor for cardiac activity. After working on the man for 45 minutes with no return of heartbeat or breathing and only pulseless electrical activity on the ECG, he was pronounced dead.

Investigators from the Coroner's Service interviewed the remaining partygoers and took photographs of the scene. Empty liquor bottles were strewn across the room and small piles of white powder were visible on a coffee table.

An autopsy of the deceased man revealed mild atherosclerosis of two coronary arteries but no other significant findings and no anatomical cause of death. Blood from the left femoral vein was submitted for toxicology with the following results.

Analysis by GC

Ethanol	150 mg/dL (33 mmol/L)

Analysis by LC-MS

Cocaine	0.8 mg/L
Benzoylecgonine	3.9 mg/L
Cocaethylene	Detected
Levamisole	Detected

- How do postmortem (after death) samples differ from samples collected in living people?
- Were the amounts of cocaine or ethanol in blood sufficient to cause death?
- What is the significance of cocaethylene and levamisole?

Discussion

Postmortem blood samples may not be collected for hours or days following death. During the interval between death and sample collection, blood does not circulate and sources of energy such as glucose are depleted. Cell membranes break down, releasing the contents of blood cells into plasma, and clots form within stationary pools of blood. These changes can affect the concentrations of many analytes, so postmortem chemistry results do not provide an accurate snapshot of the living person.

Drugs that are present in tissues may diffuse into blood, altering the concentration of drug that was present at the time of death. This phenomenon, known as postmortem redistribution, affects blood in the large vessels and heart more than peripheral blood. For this reason, blood from the femoral vein is a preferred site for sample collection.

Postmortem blood specimens are usually hemolyzed and viscous. They are not suitable for analysis by the screening immunoassays that detect drugs in urine. Instead, samples are treated with organic solvents to liberate drugs from proteins and other components of blood. The drug-containing extract is then concentrated and injected into a gas or liquid chromatograph for analysis.

When cocaine is present in a blood sample, the amount of benzoylecgonine is almost always greater than the amount of cocaine. Since benzoylecgonine is derived entirely from cocaine, adding the two concentrations gives an approximation of cocaine values at the time of death. Concentrations in excess of 0.5–1.0 mg/L are considered potentially lethal. However, there is a large overlap between lethal and nonlethal blood concentrations of cocaine/benzoylecgonine.

In this case, where no anatomical cause of death was found, the high levels of cocaine in blood provide a reasonable explanation of why the assistant director died. Ethanol is present at levels that produce intoxication, but it is well below the lethal concentration.

Cocaethylene is a toxic product formed when cocaine and ethanol are consumed together (see Figure 6.2). It is detected and reported by most forensic toxicology laboratories. Levamisole is a drug used to treat parasitic infections in animals. It is sometimes used as a cutting agent for cocaine but it has no psychoactive properties. Side-effects of levamisole include neutropenia, joint pains, and skin lesions.

Takeaway messages

- Postmortem blood is a messy specimen, and decomposition artifacts can affect the measured concentration of a drug. Despite its shortcomings, medical examiners and coroners depend upon post-mortem toxicology results to assign a cause of death.

- The toxicity of cocaine is highly variable. A concentration that might be lethal in one person is well tolerated by others.
- Street drugs are often mixed with other illicit, prescription or over-the-counter drugs before they are sold.

Case 6.3   The Incredible Hulk

Police were called to a convenience store at 2 pm in the afternoon. Upon entering, they saw a heavy-set man in his mid-30s arguing with the clerk, who was standing behind the counter holding a baseball bat.

"What's going on here?" asked one of the officers.

"This guy asked for the latest issue of Marvel comics," said the clerk. "When I told him we don't sell comic books, he went ballistic."

"You're lying," the customer barked, then jumped at the clerk and grabbed the bat.

The two officers pulled the man off the counter and tried to restrain him, but he threw them off and ran out the door yelling "I am the Incredible Hulk!"

The officers gave chase on foot and called for back-up. Three blocks from the store, they cornered the man in a blind alley.

"Easy now," said the lead officer, "let's just calm down and have a talk."

"I know you," the assailant said. "You're aliens from Alpha Centauri and you've been sent to take over the world. I can't let that happen."

He went for the policeman's service revolver but was wrestled to the ground. Four more officers arrived and managed to subdue the man and apply handcuffs. As they loaded him into a police cruiser, one of the officers noted that he was gasping for air and turning blue.

They took him to the emergency department of a nearby hospital. On admission, his vital signs were pulse 130/min, respirations 20/min, blood pressure 200/110, and temperature 103 °F (39.5 °C). His clothes were damp with perspiration.

- What is the name of this syndrome? Is it dangerous?

  A urine sample collected on admission was sent for a drug screen, with the following results.

Immunoassays

Amphetamines	Negative
Cannabis	Positive
Cocaine	Positive
Opiates	Negative

- Do these results explain the patient's clinical picture? Is additional testing necessary?

Discussion

This man is showing the signs of excited delirium. The syndrome is most frequently associated with use of cocaine, methamphetamine or another stimulant drug. Metabolic disorders, psychiatric disorders, and use or discontinuation of psychotropic drugs may also trigger the syndrome.

Excited delirium is a medical emergency that requires admission to hospital and careful monitoring. Without appropriate treatment, the mortality rate is estimated to be 5–10%.

Cocaine was detected on the urine drug screen, which explains the clinical presentation. The presence of cannabis is incidental – it is not associated with this syndrome. For purposes of managing the patient in hospital, confirmation testing is not required.

Takeaway messages

- People with excited delirium are at high risk for sudden death. They belong in a medical facility, not a jail.
- Excited delirium is diagnosed by observation and physical findings. Screening for drug use may help to identify the cause of excited delirium.

Further Reading

Article

Jones, A.W. (2019). Forensic drug profile: cocaethylene. *J. Anal. Toxicol.* 43: 155–160.

Book chapter

Prosser, J.M. and Hoffman, R.S. (2015). Cocaine. In: *Goldfrank's Toxicologic Emergencies*, 10e (eds. R.S. Hoffman, M.A. Howland, N.A. Lewin, et al.), 1054–1063. New York: McGraw-Hill.

Websites

National Institute on Drug Abuse
Cocaine
www.drugabuse.gov/publications/drugfacts/cocaine.

Centre for Addiction and Mental Health
Cocaine and Crack
www.camh.ca/en/health-info/mental-illness-and-addiction-index/cocaine.

Video

The reward circuit: how the brain responds to cocaine
www.drugabuse.gov/videos/reward-circuit-how-brain-responds-to-cocaine

7

Amphetamines

The amphetamine-type stimulants are a group of structurally related compounds that activate the central and peripheral nervous systems. These drugs have a chemical structure that is similar to the neurotransmitters dopamine and norepinephrine (Figure 7.1). Amphetamine-type stimulants are included in some over-the-counter medications, and others are available as prescription drugs.

The name amphetamine is a contraction of **alpha-m**ethyl**ph**enyl**et**hyl**amine**. The amphetamine molecule has an asymmetric alpha carbon atom (see Figure 7.1) and therefore exists as two stereoisomers, called *d* (dextrorotatory) and *l* (levorotatory). Both isomers have biological activity, but the d isomer has greater physiological effects. This difference in potency between the d and l forms applies to most of the amphetamine-type stimulants. Drug preparations may contain a mixture of both isomers, or they may consist of a single isomer (e.g., dextroamphetamine).

Modifying the structure of amphetamine alters its biological effects. Methylation of the amino group, addition of methylenedioxy, or methoxy groups to the aromatic ring, and substitutions at the alpha or beta carbon atoms produce compounds with different pharmacological properties. Structures of commonly used and abused derivatives of amphetamine appear throughout this chapter.

How Amphetamines Work

Like cocaine, amphetamines exert their stimulant effects by increasing synaptic transmission between nerve cells.

Amphetamines can enter neurons through the transporters that normally take up dopamine, norepinephrine, and serotonin from the synaptic cleft. Once inside the cell, they interfere with a vesicular monoamine transporter that packages

An Introduction to Testing for Drugs of Abuse, First Edition. William E. Schreiber.
© 2022 John Wiley & Sons Ltd. Published 2022 by John Wiley & Sons Ltd.

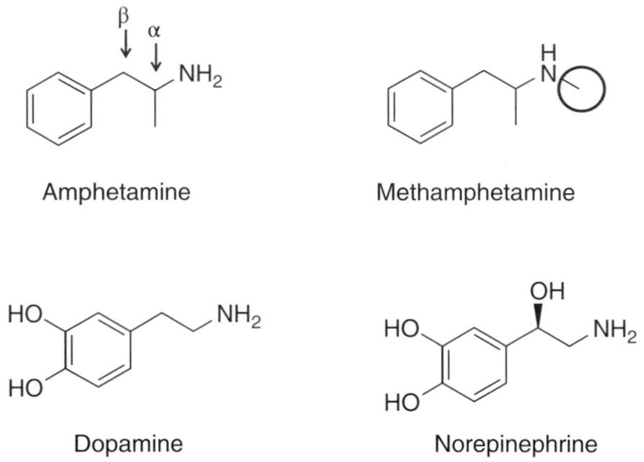

Figure 7.1 Structures of amphetamine, methamphetamine, and the neurotransmitters dopamine and norepinephrine. All of these molecules are derivatives of phenylethylamine. The alpha and beta carbon atoms of amphetamine are indicated. The circled methyl group of methamphetamine is the only difference between this compound and amphetamine.

neurotransmitters into synaptic vesicles. This depletes the vesicles of neurotransmitters and increases their concentration in the cytoplasm.

Normally, neurotransmitters are released only when a nerve impulse stimulates vesicles to fuse with the plasma membrane and release their contents. Under the influence of amphetamines, the reuptake transporters operate in reverse and neurotransmitters flow directly from cytoplasm into the synaptic cleft.

Amphetamines also interfere with the reuptake of neurotransmitters. Due to the different structures of amphetamine-type stimulants, each drug may show affinity for a particular transporter and thus potentiate the effects of that neurotransmitter. An increase in dopamine is associated with the pleasurable effects of these drugs. Increased levels of norepinephrine mediate an elevation in heart rate, blood pressure, and activity. The surge of serotonin is related to hyperthermia and hallucinations.

Physiological Effects

The major effects of amphetamines are:

- hypertension
- tachycardia
- increased body temperature, sweating
- hyperactivity, agitation

- decreased appetite
- dilated pupils
- euphoria
- hallucinations.

Some complications of amphetamine use are seizures, stroke, arrhythmias, myocardial infarction, and acute kidney injury.

Chronic users of amphetamines may develop a psychosis that resembles schizophrenia. Another long-term complication is necrotizing vasculitis. This can involve multiple organs and is responsible for the visible changes observed in the skin and gums of some patients.

Therapeutic Uses

Amphetamine and related stimulants are most commonly prescribed to treat attention deficit hyperactivity disorder (ADHD). While it may seem paradoxical to use a stimulant for this condition, amphetamines improve communication between neurons in regions of the brain responsible for attention and decision making. Amphetamine-type stimulants are also prescribed to patients with narcolepsy and obesity. Because amphetamines are addictive and have dangerous side-effects, alternative treatments are preferred for these indications. Several amphetamine-type stimulants are available in over-the-counter preparations as nasal decongestants.

Potential for Abuse

Amphetamines may be taken orally, by nasal insufflation ("snorting"), smoking, or intravenous injection. These drugs are highly addictive. Withdrawal symptoms include anxiety, fatigue, depression, and psychosis, as well as cravings for the drug.

Amphetamine and related stimulants are available by prescription and may be diverted for nonmedical use. In 2018, an estimated 5.1 million people in the United States aged 12 years and older (1.9% of this population) reported misuse of prescription stimulants in the past year. Males (2.1%) outnumbered females (1.6%), and the highest rate of misuse (6.5%) was in people aged 18–25 years.

Methamphetamine is the most commonly encountered street drug among the amphetamine-type stimulants. In 2018, more than 1.8 million people in the US aged 12 years and older (0.7% of this population) reported methamphetamine use in the past year. The prevalence was 0.8% in males and 0.5% in females, and the heaviest users were people aged 30–34 years (1.6%).

In 2017, a total of 10333 people died from drug overdoses involving psychostimulants with abuse potential (mainly methamphetamine and MDMA). This was 15% of all overdose deaths in the United States for that year.

Individual Drugs

The most commonly encountered amphetamine-type stimulants are discussed below.

Amphetamine

The prototype drug, amphetamine, was first synthesized in 1877 and was eventually marketed for its pharmaceutical properties in the 1930s. The d-isomer has 3–4 times more activity in the central nervous system (CNS) than the l-isomer. Different formulations contain the drug as the d-isomer only (Dexedrine®), a mixture of both d- and l-isomers (Adderall®), or as the prodrug lisdexamfetamine (Vyvanse®), which is converted to d-amphetamine following absorption.

Amphetamine is widely prescribed to treat ADHD. It may also be prescribed for narcolepsy and on a short-term basis for weight reduction. The half-life is 7–34 hours. Amphetamine is metabolized by the liver to a number of compounds, but about 30% of the drug is excreted unchanged. Clinical assays for amphetamine detect only the parent drug, not its metabolites.

Methamphetamine

Methamphetamine, the N-methyl derivative of amphetamine, is a common drug of abuse (Figure 7.1). Most of the methamphetamine consumed in North America is made in clandestine drug labs and sold through illicit channels. This was portrayed to great dramatic effect in the popular television series Breaking Bad. The lead character was a high school chemistry teacher who turned to synthesizing methamphetamine as a way to increase his income and, eventually, his personal standing in the world.

The d-isomer of methamphetamine has much greater CNS activity than the l-isomer. d-Methamphetamine is available by prescription (Desoxyn®) for treatment of ADHD and obesity. Because of its high abuse potential, other medications are usually preferred. The l-isomer is an effective vasoconstrictor and is available over the counter as a nasal decongestant (Vicks inhaler).

The half-life of d-methamphetamine is 6–15 hours. About 40% of the drug is excreted unchanged, and smaller amounts are metabolized to amphetamine. Both compounds are detected in urine or blood following use of methamphetamine.

Illicit Drugs

Methylenedioxymethamphetamine (MDMA)

MDMA is a derivative of methamphetamine in which a methylenedioxy group has been added to the phenyl ring (Figure 7.2). It has a particular affinity for the serotonin transporter and causes excessive release of serotonin from nerve endings. People taking MDMA are especially prone to hyperthermia and its consequences.

The drug, which goes by the street name ecstasy, became popular for its use at all-night dance parties. It creates feelings of trust and closeness and promotes social interactions. The terms "empathogen" and "enactogen" are sometimes applied to drugs with these properties. MDMA is usually taken as a tablet or capsule, and the effects last 3–6 hours.

Methylenedioxyethylamphetamine (MDEA), which has an ethyl group attached to the amine, produces similar effects to MDMA. However, those effects are milder and do not last as long.

The half-life of MDMA is 5–9 hours. It is metabolized to methylenedioxyamphetamine (MDA), which has biological activity, as well as other less active metabolites. Both MDMA and MDA are excreted in urine, where they can be detected and measured. MDEA is metabolized and excreted in a similar manner.

Paramethoxyamphetamine (PMA) and
Paramethoxymethamphetamine (PMMA)

PMA and PMMA are derivatives of amphetamine and methamphetamine, respectively, in which a methoxy group is attached to the para position of the

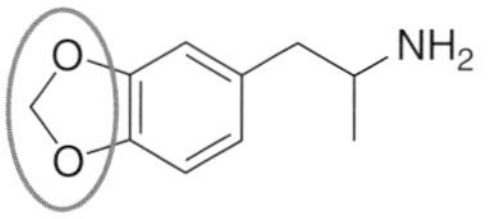

Methylenedioxyamphetamine
(MDA)

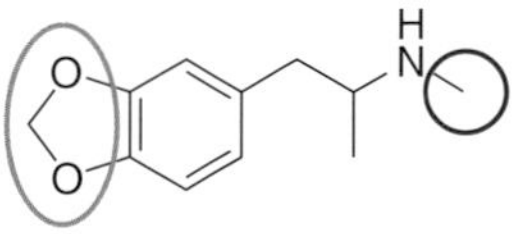

Methylenedioxymethamphetamine
(MDMA)

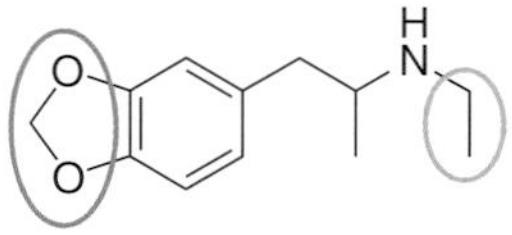

Methylenedioxyethylamphetamine
(MDEA)

Figure 7.2 Structures of methylenedioxyamphetamine (MDA), methylenedioxymethamphetamine (MDMA), and methylenedioxyethylamphetamine (MDEA). Differences between amphetamine and each of these compounds are highlighted with circles and ellipses.

Paramethoxyamphetamine
(PMA)

Paramethoxymethamphetamine
(PMMA)

Figure 7.3 Structures of p-methoxyamphetamine (PMA) and
p-methoxymethamphetamine (PMMA). Differences between amphetamine and each of
these compounds are highlighted with circles and ellipses.

phenyl ring (Figure 7.3). They have similar effects to MDMA but are more potent, and their ingestion has been linked to deaths in several countries. These drugs may be taken as a substitute for MDMA, or they may be included in street drugs which are sold as MDMA.

PMMA is metabolized to PMA, and both compounds are further transformed to other metabolites. Detection of both PMMA and PMA in a urine specimen indicates use of PMMA, while detection of only PMA is seen with ingestions of that drug.

Cathinones

Cathinone is a natural product found in the khat plant (*Catha edulis*), which grows in eastern Africa and southern Arabia. People who chew the leaves experience a mild stimulant effect. It has the same structure as amphetamine except that the beta carbon atom has a ketone group (Figure 7.4).

Cathinone

Methcathinone

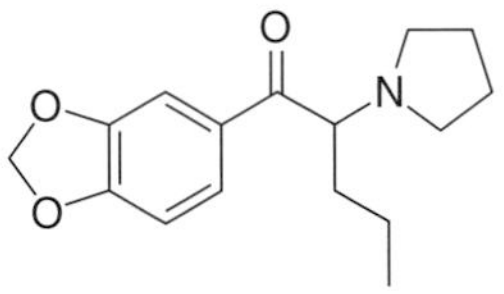

Methylenedioxypyrovalerone
(MDPV)

Figure 7.4 Structures of cathinone and the synthetic derivatives methcathinone and methylenedioxypyrovalerone (MDPV). Differences between amphetamine, cathinone, and methcathinone are highlighted with circles and ellipses.

Synthetic cathinones are more powerful than the naturally occurring compound. They are consumed in the same manner as other amphetamine-type stimulants and can be toxic or fatal. These compounds have been labeled and sold as "bath salts" because of their crystalline appearance.

There are many different synthetic cathinones. Two examples, methcathinone and methylenedioxypyrovalerone (MDPV), are shown in Figure 7.4. They are not detected by the usual screening immunoassays, but they can be identified by labs that have developed specific mass spectrometry-based analyses.

Over-the-Counter and Prescription Drugs

Pseudoephedrine/Ephedrine

These two compounds have the same structural formula (Figure 7.5), but they differ in the spatial arrangement of groups at the alpha and beta carbon atoms. The d-isomer of pseudoephedrine and the l-isomer of ephedrine are found in species of the *Ephedra* plant. Herbal preparations from this plant may be consumed to promote energy or weight loss.

Pseudoephedrine is available over the counter as a nasal decongestant. Its availability has become more restricted, because it can be used as a starting material for the synthesis of methamphetamine. Ephedrine is a decongestant and bronchodilator that may be prescribed to relieve shortness of breath or wheezing due to asthma.

The half-life of pseudoephedrine is 3–16 hours, and the half-life of ephedrine is 4–10 hours. Both compounds are excreted mostly unchanged in the urine, and they may cross-react in immunoassays for amphetamines.

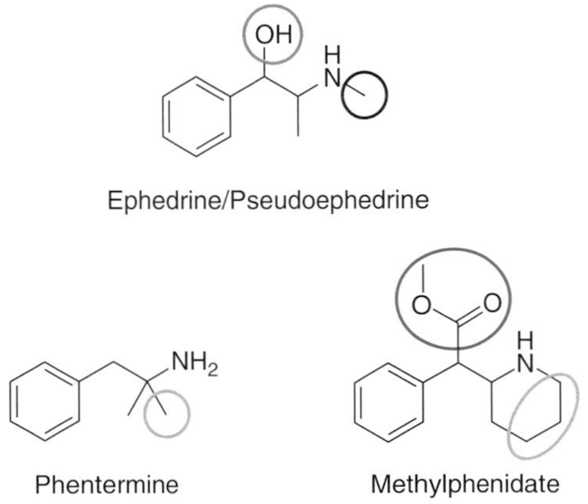

Figure 7.5 Structures of ephedrine/pseudoephedrine, phentermine, and methylphenidate. Pseudoephedrine and ephedrine have the same structural formula but differ in the orientation of substituents on the alpha and beta carbon atoms. Differences between amphetamine and each of these compounds are highlighted with circles and ellipses.

Ephedrine/Pseudoephedrine

Phentermine

Methylphenidate

Methylphenidate

Methylphenidate is prescribed for the treatment of attention deficit disorder, ADHD, and narcolepsy. Similar to amphetamine, the d-isomer has greater activity than the l-isomer, and formulations of the d-isomer (dexmethylphenidate) are commercially available.

The half-life is about 1–4 hours. Most of the drug is metabolized to ritalinic acid and excreted in urine, where it can be detected by mass spectrometry-based techniques.

Phentermine

Phentermine, the alpha-methyl derivative of amphetamine, is prescribed as an appetite suppressant to treat obesity. It has a half-life of 19–24 hours, and most of the drug is excreted unchanged in urine. Due to its similarity with amphetamine, it can cause false-positive results in screening immunoassays.

Testing for Amphetamines

Screening

Immunoassays are the usual front-line tests for amphetamines. The assay may be configured to detect amphetamine and methamphetamine only, or it may detect a broader range of amphetamine-type stimulants, including illicit drugs (e.g., MDMA) and over-the-counter stimulants (e.g., pseudoephedrine). Some assays discriminate between the d- and l-isomers of these compounds. Table 7.1 lists amphetamine-type stimulants that are detected in one commercial assay.

According to SAMHSA guidelines, the cut-off for a positive screening test is 500 ng/mL. Confirmatory testing is considered positive at a concentration of 250 ng/mL or higher. To report a positive result for methamphetamine, the concentration of amphetamine must be at least 100 ng/mL.

Whether positive or negative, screening tests for amphetamines can be difficult to interpret. A positive test is evidence of an amphetamine, but many compounds can cross-react in immunoassays. A negative test does not rule out an amphetamine-type stimulant, since some of these drugs (e.g., cathinones) react weakly or not at all in immunoassays.

Confirmation

To confirm a positive screening result, samples can be analyzed by gas chromatography-mass spectrometry (GC-MS) or liquid chromatography-mass spectrometry (LC-MS). Amphetamine, methamphetamine, MDMA, and MDA are

Table 7.1 Cross-reactivity of amphetamine-type stimulants in a commercial immunoassay.

Compound	% Cross-reactivity
d-Amphetamine	104
l-Amphetamine	1.0
d-Methamphetamine	100
l-Methamphetamine	18
3,4-Methylenedioxyamphetamine (MDA)	116
3,4-Methylenedioxymethamphetamine (MDMA)	196
3,4-Methylenedioxyethylamphetamine (MDEA)	172
p-Methoxyamphetamine (PMA)	24
p-Methoxymethamphetamine (PMMA)	100
Phentermine	3.3
d-Pseudoephedrine	0.9
l-Ephedrine	0.5

The assay is calibrated to detect d-methamphetamine at the specified cut-off value. Compounds with >100% cross-reactivity give a positive result at lower concentrations; compounds with <100% cross-reactivity give a positive result at higher concentrations.
Source: CEDIA Amphetamine/Ecstasy Assay, Microgenics Corporation.

routinely detected and measured by most toxicology laboratories. Assays for other amphetamine-type stimulants are available through regional or national reference laboratories.

One weakness of these confirmation methods is that d- and l-isomers are not resolved by mass spectrometry. If methamphetamine is confirmed in a patient sample, it could be from an illicit source (d-isomer) or an over-the-counter decongestant (l-isomer). Special columns that can resolve the d- and l- forms of amphetamines by chromatography are used by some labs. Your local toxicologist can provide more information about which compounds are detected.

Window of Detection

Amphetamine and methamphetamine can be detected in urine samples up to 3 days after drug ingestion. MDMA and MDA are detected for up to 2 days. These time periods are approximate and depend upon several factors: (i) frequency and amount of drug use, (ii) urine pH, which affects the rate of drug excretion, and

(iii) the detection limit used by the laboratory (which may be lower than the SAMHSA cut-off).

Interferences

Many prescription drugs can give a positive screening result for amphetamines (Table 7.2). In some cases, a precursor drug (e.g., selegiline, benzphetamine) is metabolized to amphetamine or methamphetamine. In other cases, an amphetamine-type stimulant (e.g., phentermine) is present. A number of drugs are not related to the amphetamines but still react in the screening test.

To properly interpret the results of a positive screening test, the patient's complete list of medications should be reviewed (prescription and over the counter). Confirmation testing will identify or exclude the most commonly used amphetamines. Additional testing for other amphetamines may need to be sent to a reference toxicology laboratory.

Table 7.2   Drugs that can give a positive test result in an immunoassay for amphetamines.

Amantadine	Methylphenidate
Aripiprazole	Methamphetamine
Atomoxetine	MDMA
Benzphetamine	Phentermine
Bupropion	Promethazine
Clobenzorex	Pseudoephedrine
Chlorpromazine	Phenylephrine
Desipramine	Phenylpropanolamine
Dextroamphetamine	Ranitidine
Dimethylamylamine	Ritodrine
Ephedrine	Selegiline
Fenproporex	Thioridazine
Isometheptene	Trazodone
Isoxsuprine	Trimipramine
Labetalol	Trimethobenzamide
Metformin	

Source: Moeller KE, Kissack JC, Atayee RS, et al. Clinical interpretation of urine drug tests: what clinicians need to know about urine drug screens. Mayo Clin Proc 2017;92:774–796.

Case Studies

Case 7.1 An Unexpected Result

Following a lengthy application process and multiple interviews, a 38-year-old woman was offered a position as an air traffic controller, which she happily accepted. Before starting her new job, she was required to submit a urine specimen for a drug screen. Five days later, she was contacted by Dr Checkup, who identified himself as a medical review officer (MRO). He advised the woman that her urine sample had tested positive for an illicit drug, and he asked if there was a reason for this finding.

The laboratory report contained the following information.

Immunoassay

Amphetamines	Positive
Cannabis	Negative
Cocaine	Negative
Opiates	Negative
Phencyclidine	Negative

Analysis by LC-MS

Amphetamine	820 ng/mL
Methamphetamine	1900 ng/mL

- What is a MRO?
- Do these results indicate that the woman took methamphetamine?
- Could these test results be due to use of illicit drugs? Prescription drugs?

Discussion
A MRO is a licensed physician who reviews laboratory results from employer drug testing programs. The MRO determines if there is a legitimate medical explanation for positive test results and for samples reported as adulterated, substituted, or invalid. The MRO functions as an impartial advocate for the accuracy and integrity of the drug testing process.

The presence of methamphetamine and amphetamine in this woman's urine sample indicates that she consumed either methamphetamine or another drug that is metabolized to methamphetamine. The source of the drugs cannot be determined from test results alone. It is possible that she bought the methamphetamine from a street dealer, or that she filled a prescription at a local pharmacy.

On her job application, she neglected to mention that she attended a weight loss clinic. Her treatment program included a 4-week course of the appetite suppressant benzphetamine, 50 mg/d. Benzphetamine is metabolized to methamphetamine, which in turn is partially metabolized to amphetamine. She was 2 weeks into the program when her preemployment urine sample was collected. Therefore, the urine screening test was positive for amphetamines and their identity was confirmed by LC-MS.

Upon hearing about the benzphetamine prescription, Dr Checkup said that explained the positive methamphetamine result. He notified the employer that the urine drug test was negative, and she started her new job the following week.

Takeaway messages

- Some prescription medications are metabolized to compounds that are considered drugs of abuse. They will be detected by standard screening and confirmation tests.
- Both methamphetamine and drugs that are metabolized to methamphetamine are infrequently prescribed. When present, methamphetamine is usually from an illicit source.
- In this case, a review of the applicant's medications provided a reason for her test results.

Case 7.2   Party Time at The Adrenaline

Emergency medical services responded to a call from the Adrenaline Club, a local nightclub for young people, at 1 am. On arrival, the paramedics were led to an unconscious, naked man in his mid-twenties lying in front of a microphone stand. According to witnesses, the victim was the lead singer of the band that was performing that evening. Twenty minutes earlier, he had begun to act in a bizarre fashion, yelling obscenities into the microphone and removing his clothes before vomiting and then collapsing onto the stage.

During transport to the hospital, his vital signs showed a pulse of 120 beats per minute, blood pressure of 180/110 and a temperature of 102 °F (39 °C). The emergency room physician suspected a cocaine overdose and asked for a stat drug screen on a catheterized urine sample. Test results came back an hour later:

Immunoassay

Amphetamines	Negative
Benzodiazepines	Negative
Cannabis	Negative
Cocaine	Negative
Opiates	Negative

- Do these results rule out a drug intoxication?

  By 6 a.m., the patient had regained consciousness and his vital signs had normalized. When questioned about whether he had taken any drugs earlier that evening, he said yes, he bought ecstasy from one of the nightclub bouncers and took it just before going on stage.

- Does ecstasy (MDMA) show up in urine drug screens for amphetamines?

  Before leaving the hospital, the patient handed a small packet of crystals to the attending physician and said, "I'm never taking this stuff again!"

Discussion

A battery of negative screening tests does not rule out a drug ingestion, because screening assays only detect a limited number of drugs. The negative result for cocaine is very reliable, but there are many amphetamine-like compounds that are not picked up by screening assays. Clinicians need to discuss which drugs are detected with the laboratory staff to ensure that a negative result is interpreted correctly.

The emergency doctor conferred with the pathologist on call and confirmed that MDMA and its metabolite MDA were both detected in the screening assay. Given the negative screening results for cocaine and amphetamines, the doctor suspected that another amphetamine-like substance, such as cathinones (bath salts), might be responsible for the patient's strange behavior and collapse. After discharging the patient, he sent the urine sample and the packet of crystals to a specialty toxicology laboratory and asked for a complete stimulant profile.

One week later, he received the report – MDPV (Figure 7.4) was identified as the main component of the crystals and was also present in the patient's urine sample. He called the pathologist to ask if MDPV can be detected in the amphetamines assay – it is not detected.

Takeaway messages

- When screening tests are negative but suspicion of a drug ingestion is high, review the performance of those tests to determine which drugs can be excluded.

- Further investigation may require the services of a specialized laboratory and will not be available in time to manage the acute in-hospital episode.
- The actual composition of street drugs may be different from what is advertised. In this case, the patient thought he was buying ecstasy and instead purchased MDPV (a cathinone derivative).

Case 7.3   Angry Doctor

A 23-year-old male who works in the food service industry was prescribed methylphenidate, 20 mg/d, for ADHD. At his next clinic visit, he reported that his ability to concentrate had improved, and he was performing better at his job. The physician asked him for a urine sample to check on whether he was taking the medication as directed. An in-office test of the urine sample was negative for amphetamines.

The doctor confronted the patient and asked him why he was not taking the drug as prescribed. When the patient protested and said he was doing exactly as instructed, the doctor accused him of lying.

- Was the doctor's reaction justified? Why?
- What advice would you give to this doctor?

Discussion

Methylphenidate is one of several amphetamine-type stimulants used to treat ADHD. The physician may have assumed that his point-of-care drug test would detect methylphenidate because other medications used to treat ADHD, such as dextroamphetamine and lisdexamfetamine, give a positive result. However, its structure is different enough from amphetamine (Figure 7.5) that it is not detected by most immunoassay-based screening tests for amphetamines.

The doctor's reaction was based on an incorrect assumption and was therefore inappropriate. To check for compliance, he should have sent the urine sample to a toxicology laboratory and requested analysis for methylphenidate or its metabolite ritalinic acid.

Takeaway messages

- Healthcare practitioners who use point-of-care drug tests in their clinics need to be familiar with which drugs are detected and which ones are not.
- As a rule, point-of-care tests are not as reliable as tests performed in a central laboratory. If unexpected results are obtained, it is best to confirm those results at a reference laboratory before acting on them.

Case 7.4   On the Road Again

JB, a 49-year-old man, works as a long-haul truck driver for Big Daddy's Transport. The company tests its drivers intermittently for drug use to ensure that they are not driving under the influence. One day, as JB was barreling down the interstate singing "On the Road Again" by Willie Nelson, he received a call on his cell phone.

"Jimbo, this is Big Daddy at headquarters. We just heard from the lab that your last urine sample was hot. After you drop your load, get back to HQ and report to me pronto."

When JB returned to the company office, he was handed a test report with the following results.

Immunoassay

Amphetamines	Positive
Cannabis	Negative
Cocaine	Negative
Opiates	Negative
Phencyclidine	Negative

"This can't be right," he said. "I don't use speed or any of that junk."

"Sorry, Jimbo," Big Daddy responded. "You know the rules. You're suspended until further notice."

- How common are positive drug tests in the trucking industry?
- Did JB take amphetamines? What other drugs or substances could have caused the positive screening test?
- Were correct procedures followed in reporting JB's test results?

Discussion

Trucking companies are required to test their drivers for a panel of five drugs on a random basis throughout the year. The Federal Motor Carrier Safety Administration, which regulates the industry, estimates that the positive test rate for commercial drivers is about 1%. More than half of the positive tests are due to cannabis. Cocaine is the next most commonly detected drug, followed by methamphetamine and amphetamine.

In order to report a positive result, the concentration of drug in the urine sample must exceed the specified cut-offs in both an initial test (usually an immunoassay) and a confirmatory test (typically GC-MS or LC-MS). Many drugs can give a positive result for amphetamines in a screening immunoassay (see Table 7.2). The list includes prescribed medications, over-the-counter remedies, and illicit substances. JB's test results are presumptive only and need to be confirmed by a definitive method before concluding that he did indeed take amphetamines.

JB had been taking a cold preparation containing pseudoephedrine at the time his urine sample was collected. The structure of pseudoephedrine is similar to amphetamine (see Figure 7.5), and some immunoassays give a positive result when pseudoephedrine is present in urine.

Unfortunately, correct procedures were not followed by the laboratory. A member of the lab staff inadvertently released the result of the initial test to the employer, rather than waiting for the confirmation test to be performed. Analysis of the urine sample by LC-MS detected no amphetamine, methamphetamine or MDMA, and the test was eventually reported as negative.

Three days later, Big Daddy called JB on his cell phone again.

"Hey Jimbo, listen up. I just got another test result, and this time it's negative. Maybe I jumped the gun when I sacked you."

"That's what I tried to tell you," JB said. "The only thing I've been taking is cold medicine. It's perfectly legal, and it helps me to stay awake."

Takeaway messages

- Screening tests for amphetamines are subject to false-positive results from a number of over-the-counter and prescription drugs. A positive screening result must be confirmed by GC-MS or LC-MS before making a judgment about amphetamine use.
- Pseudoephedrine is included in many cold preparations and is a common source of false-positive results.
- Employers need to be familiar with the drug testing process and follow accepted procedures when dealing with their employees.

Further Reading

Articles

Liu, L., Wheeler, S.E., Venkataramanan, R. et al. (2018). Newly emerging drugs of abuse and their detection methods. *Am. J. Clin. Pathol.* 149: 105–116.

Prosser, J.M. and Nelson, L.S. (2012). The toxicology of bath salts: a review of synthetic cathinones. *J. Med. Toxicol.* 8: 33–42.

Book chapter

Jang, D.H. (2015). Amphetamines. In: *Goldfrank's Toxicologic Emergencies*, 10e (eds. R.S. Hoffman, M.A. Howland, N.A. Lewin, et al.), 1030–1041. New York: McGraw-Hill.

Websites

National Institute on Drug Abuse

Methamphetamine
www.drugabuse.gov/publications/drugfacts/methamphetamine
MDMA (Ecstasy/Molly)
www.drugabuse.gov/publications/drugfacts/mdma-ecstasymolly
Prescription stimulants
www.drugabuse.gov/publications/drugfacts/prescription-stimulants
Synthetic cathinones (bath salts)
www.drugabuse.gov/publications/drugfacts/synthetic-cathinones-bath-salts

Centre for Addiction and Mental Health

Amphetamines
www.camh.ca/en/health-info/mental-illness-and-addiction-index/amphetamines
Ecstasy
www.camh.ca/en/health-info/mental-illness-and-addiction-index/ecstasy
Methamphetamines
www.camh.ca/en/health-info/mental-illness-and-addiction-index/
    methamphetamines

Videos

The reward circuit: how the brain responds to methamphetamine.
www.drugabuse.gov/videos/reward-circuit-how-brain-responds-
    to-methamphetamine

AACC Pearls of Laboratory Medicine

Synthetic drugs: cathinones and cannabinoids
www.aacc.org/science-and-research/clinical-chemistry-trainee-council/trainee-
    council-in-english/pearls-of-laboratory-medicine/2015/synthetic-drugs-
    cathinones-and-cannabinoids

8

Benzodiazepines and Z-drugs

Benzodiazepines

The benzodiazepines are a class of sedative-hypnotic drugs. Their chemical structure contains a **benz**ene ring fused to a **diazepine** ring, for which the compounds are named. A phenyl group is attached to the diazepine ring, producing the core structure that is common to all benzodiazepines.

In the early 1960s, chlordiazepoxide and diazepam were the first benzodiazepines to be introduced into clinical practice (Figure 8.1). Numerous analogs have been synthesized, and a number of these are in clinical use today (Table 8.1 and Figures 8.2–8.4).

How Benzodiazepines Work

Benzodiazepines depress the activity of the central nervous system (CNS) by enhancing the effects of gamma-aminobutyric acid (GABA).

GABA is a neurotransmitter released by inhibitory neurons within the CNS. Inhibitory neurons form synapses with the cell bodies, dendrites, and axons of excitatory neurons and restrict the transmission of nerve impulses within the CNS.

When an inhibitory neuron fires, it releases GABA from the presynaptic nerve ending. After crossing the synaptic cleft, GABA binds to the GABA-A receptor, a chloride transporter located on the postsynaptic membrane of an adjoining neuron. Upon binding to GABA, the chloride channel opens and chloride ions enter the cell. The influx of chloride increases the negative charge within the neuron, which becomes hyperpolarized, and prevents it from firing.

An Introduction to Testing for Drugs of Abuse, First Edition. William E. Schreiber.
© 2022 John Wiley & Sons Ltd. Published 2022 by John Wiley & Sons Ltd.

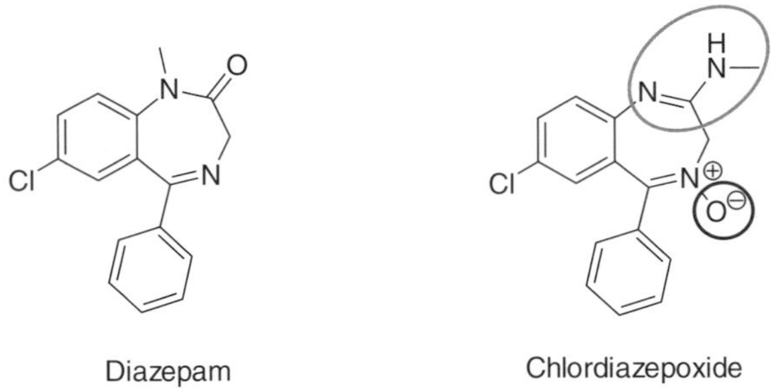

Diazepam   Chlordiazepoxide

Figure 8.1 Structures of diazepam and chlordiazepoxide. The differences between diazepam and other benzodiazepines are indicated by circles and ellipses in Figures 8.1–8.4.

Table 8.1 Potency, half-life, and clinical uses of common benzodiazepines.

Drug	Equivalent dose (mg)	Half-life (h)	Clinical use
Alprazolam	1.0	6–27	Anxiety, panic disorder
Chlordiazepoxide	30	6–27	Anxiety, alcohol withdrawal
Clonazepam	1.0	19–60	Seizures, panic disorder
Clorazepate	15	31–97[a]	Anxiety, seizures, alcohol withdrawal
Diazepam	10	21–37	Anxiety, alcohol withdrawal, seizures, muscle spasms
Flurazepam	30	1–3	Insomnia
Lorazepam	1–2	9–16	Anxiety, insomnia, seizures
Oxazepam	20	4–11	Anxiety, alcohol withdrawal
Temazepam	20	3–13	Insomnia
Triazolam	0.5	1.8–3.9	Insomnia

[a] As nordiazepam.
Source: Data are from multiple sources.

Benzodiazepines bind to the GABA-A receptor. They do not open the chloride channel in the absence of GABA, but when GABA is bound to the receptor, they increase the frequency of channel opening. This allows more chloride ions to flow into the neuron and augments the inhibitory effect of GABA. A second receptor, GABA-B, is also found in the CNS, but it is not activated by benzodiazepines.

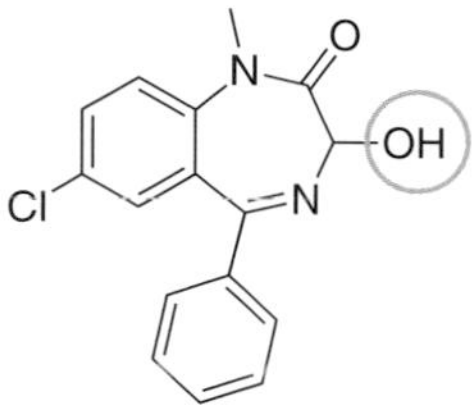

Temazepam

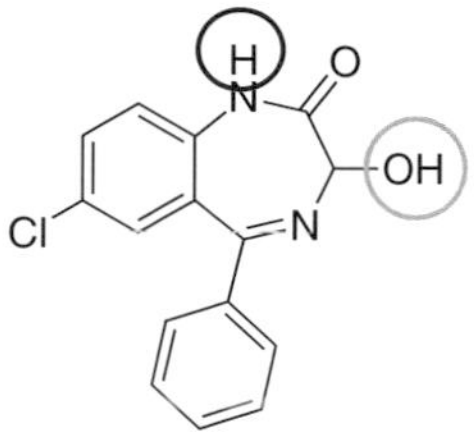

Oxazepam

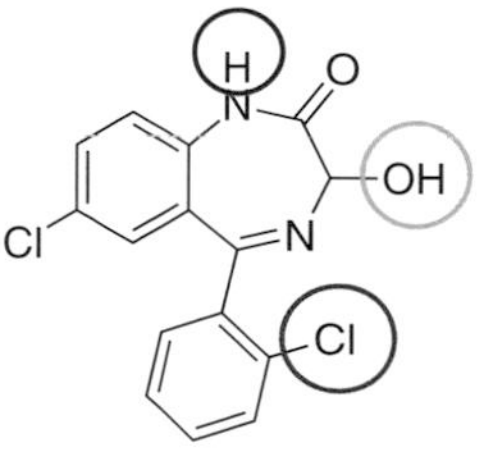

Lorazepam

Figure 8.2 Structures of temazepam, oxazepam, and lorazepam. These drugs are conjugated to glucuronic acid through the OH group prior to excretion.

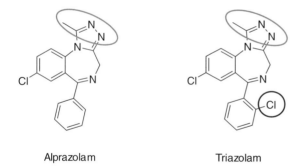

Alprazolam

Triazolam

Figure 8.3 Structures of alprazolam and triazolam.

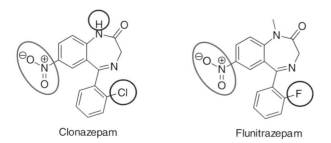

Clonazepam

Flunitrazepam

Figure 8.4 Structures of clonazepam and flunitrazepam. Flunitrazepam has been used in cases of sexual assault to induce drowsiness and amnesia. It is not approved for medical use in the United States and Canada.

Physiological Effects

Benzodiazepines produce the following effects.

- Sedation
- Reduction of anxiety
- Hypnosis (sleep)
- Anticonvulsant activity
- Muscle relaxation
- Amnesia
- Dizziness
- Loss of coordination
- Inability to concentrate

Some patients have a paradoxical reaction to these drugs and may become restless, agitated, aggressive, or even violent.

Therapeutic Uses

- *Anxiety states* – benzodiazepines may be prescribed on a short-term basis for situational anxiety (i.e., stressful events) or longer term for generalized anxiety disorder and panic disorders.
- *Insomnia* – benzodiazepines with a short half-life are often prescribed as sleeping aids.
- *Medical procedures* – drugs such as midazolam are given before medical or surgical procedures to provide sedation and relieve anxiety.
- *Seizures* – acute seizures can be controlled with several of the benzodiazepines.
- *Alcohol withdrawal* – longer-acting benzodiazepines are used to treat patients undergoing alcohol withdrawal.
- *Muscle spasms* – diazepam is effective for treating muscle spasticity that originates in the CNS.

From 2003 to 2015, the rate of prescribing benzodiazepines among primary care physicians in the United States more than doubled.

Potential for Abuse

Benzodiazepines are safer than barbiturates and have largely replaced them as sedative-hypnotic agents. However, tolerance and dependence can occur with this class of drugs to an equivalent or greater degree than opioids. Symptoms of benzodiazepine withdrawal include anxiety, restlessness, agitation, insomnia, hyperactive reflexes, and seizures. Drugs with a longer half-life produce less severe symptoms, as they are eliminated more slowly from the body.

In 2018, an estimated 5.4 million people in the US aged 12 years and older (2.0% of this population) reported misuse of prescription benzodiazepines in the past year. Rates of misuse were about the same for males and females and were highest in people aged 18–25 years (4.5% of this age cohort).

More than one million people aged 12 years and older (0.4% of this population) reported misuse of prescription sedatives in 2018. This includes the "Z-drugs" zolpidem, zaleplon, and eszopiclone as well as several short-acting benzodiazepines (e.g., flurazepam, temazepam, triazolam) and barbiturates. Misuse was somewhat higher among males than females, and the highest rate of misuse was in people aged 18–39 years (0.6% of this age group).

Because of their safety profile, benzodiazepines are rarely responsible for deaths on their own. However, when taken in combination with opioids or ethanol, they contribute to respiratory depression and can be fatal. In 2017, 11 537 deaths in the United States involved benzodiazepines. In the majority of cases, an opioid was also present.

Flumazenil is a benzodiazepine analog that binds to the GABA-A receptor and blocks the actions of benzodiazepines. It can be used to treat overdoses or to reverse the effects of benzodiazepines that are given during medical procedures.

Metabolism

The major metabolites of some common benzodiazepines are shown in Figure 8.5. Most of these metabolites are conjugated to glucuronic acid, either directly or following additional chemical modification, and excreted in urine. Temazepam, oxazepam, and lorazepam are excreted as the glucuronide conjugates. Only small amounts of the other parent drugs, either free or conjugated, appear in urine.

The half-lives of the benzodiazepines are given in Table 8.1. Many of the major metabolites have biological activity, and in some cases they are important contributors to the drug's effects. They also influence the time during which the drug can be detected.

Testing for Benzodiazepines

Screening

Screening immunoassays for benzodiazepines can detect most of the commonly prescribed drugs in this group (Table 8.2). The cut-off used by clinical laboratories is usually 200 or 300 ng/mL. A positive result indicates the presence of one or more benzodiazepines, but it does not indicate which ones are present. For monitoring patients on a prescription, a positive screening test is ordinarily sufficient evidence of compliance.

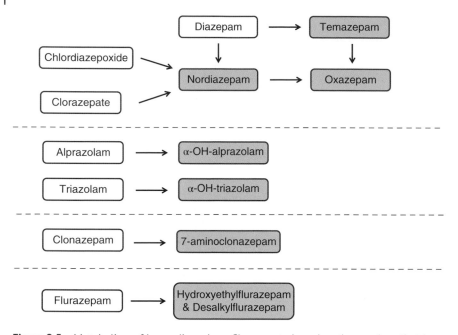

Figure 8.5 Metabolism of benzodiazepines. Clorazepate is an inactive prodrug that is converted by gastric acid into the active metabolite nordiazepam. Temazepam and oxazepam may be parent drugs or metabolites of another benzodiazepine. Lorazepam (not pictured) does not produce metabolites.

Table 8.2 Cross-reactivity of selected benzodiazepines in a commercial immunoassay.

Benzodiazepine	Metabolite	% Cross-reactivity
Alprazolam		205
	alpha-OH-Alprazolam	188
Chlordiazepoxide		13
	Nordiazepam	211
Clonazepam		140
	7-Aminoclonazepam	96
Diazepam		247
Flurazepam		189
	Desalkylflurazepam	210
Lorazepam		122
Oxazepam		107
Temazepam		144
Triazolam		191
	alpha-OH-Triazolam	193

Source: CEDIA Benzodiazepine Assay, Microgenics Corporation.

False-negative screening tests for benzodiazepines may occur for several reasons. First, most of these drugs and their metabolites are conjugated to glucuronic acid before excretion. Immunoassays may detect the free drug/metabolite but not the conjugate. Treatment with glucuronidase prior to analysis increases the likelihood of detecting benzodiazepines and their metabolites.

Second, some benzodiazepines may test negative in the screening assay because their concentration is below the detection threshold. Alprazolam, clonazepam, lorazepam, and triazolam are all taken in smaller doses and may evade detection (Table 8.1). Some researchers have proposed a lower cut-off to improve screening test performance.

Confirmation

Confirmation testing can identify individual benzodiazepines and measure their concentrations. Both gas chromatography-mass spectrometry (GC-MS) and liquid chromatography-tandem mass spectrometry (LC-MS/MS) methods are in widespread use. Sensitivity is higher than for the screening test. The menu of drugs that can be detected and the cut-off values for a positive result vary between laboratories.

Lorazepam, temazepam, and oxazepam are identified as the parent drug. Other benzodiazepines are extensively transformed, and only the metabolites may appear in urine (Figure 8.5). For some drugs (e.g., diazepam), more than one metabolite is present.

Window of Detection

Most benzodiazepines can be detected in urine samples up to 5 days after ingestion. Diazepam can be detected up to 10 days, while the short-acting drug triazolam disappears after 1–2 days. These times are approximate and depend upon the dose and for how long the drug has been taken. With chronic use, some benzodiazepines may be detected for up to 30 days after the most recent dose.

Interferences

Screening assays are very specific, and only a few drugs have been reported to give false-positive results. They include oxaprozin (a nonsteroidal antiinflammatory drug), efavirenz (a treatment for HIV), and the antidepressant medication sertraline.

Z-Drugs

Zaleplon, zolpidem, and zopiclone are all short-acting hypnotic drugs. These compounds are often referred to as "Z-drugs" because they begin with the letter Z and have similar pharmacological properties. They are structurally distinct from the benzodiazepines and other sedative-hypnotic drugs (Figure 8.6).

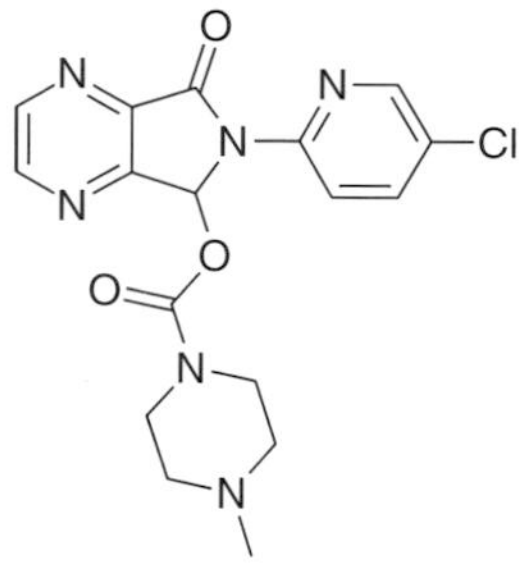

Zopiclone

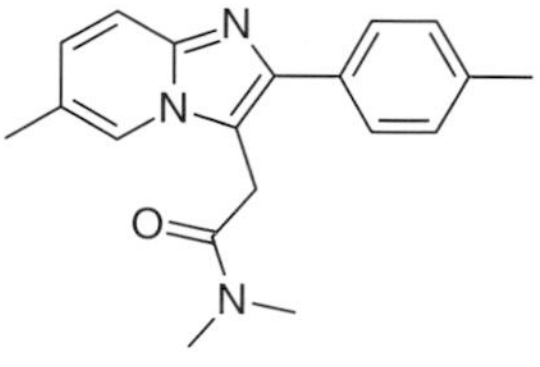

Zolpidem

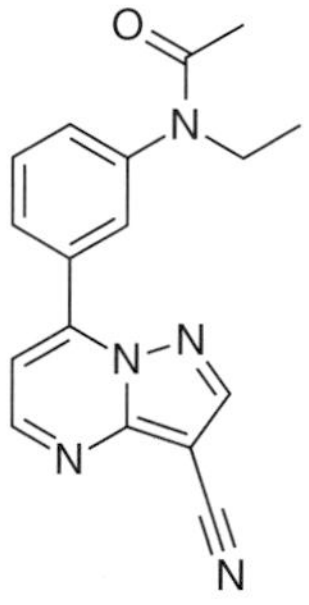

Zaleplon

Figure 8.6 Structures of the Z-drugs.

The first Z-drug to be approved in the US was zolpidem in 1992, followed by zaleplon in 1999 and eszopiclone, the S-isomer of zopiclone, in 2004.

How Z-Drugs Work

All three drugs bind to the benzodiazepine site on the GABA-A receptor and promote chloride channel opening. They have the same sleep-producing effects as benzodiazepines, but without the anxiolytic or muscle relaxing properties.

Therapeutic Uses

The Z-drugs are prescribed to treat insomnia. Their rapid onset of action and short half-lives help people to fall and remain asleep without causing drowsiness the next day.

Potential for Abuse

The numbers of people who misuse Z-drugs and other hypnotics is given in the section on benzodiazepines. Overdoses can be fatal, particularly when alcohol or other sedating medications are coingested.

Table 8.3 Dosage and half-life of Z-drugs.

Drug	Nightly dose (mg)	Half-life (h)
Zaleplon	5–20	0.9–1.2
Zolpidem	5–10	1.4–4.5
Eszopiclone[a]	1–3	4–9

[a] S-isomer of zopiclone.
Source: Data are from multiple sources.

Metabolism

All the Z-drugs are metabolized to other compounds. The major excretion forms in urine are (i) 5-oxozaleplon, (ii) zolpidem phenyl-4-carboxylic acid, and (iii) N-desmethylzopiclone. Less than 5% of the parent drugs appears in urine.

Testing for Z-Drugs

Most clinical laboratories do not include Z-drugs in their screening panels for drugs of abuse. Analysis of these compounds is usually performed by gas or liquid chromatography using an MS or MS/MS detector. Methods that detect only the parent compounds are less sensitive than assays that identify metabolites as well.

Window of Detection

Because their half-lives are short when compared to benzodiazepines (Table 8.3), the Z-drugs can be detected for up to 3 days after ingestion.

Case Studies

Case 8.1   Nervous, Anxious, and Worried

A 24-year-old woman made an appointment to see her family doctor for chronic nervousness. During the visit, she admitted to a paralyzing fear of making the wrong decisions, worrying about the health of family members, and inability to relax for more than a few minutes at a time. She did not sleep well at night, felt fatigued during the day, and had intermittent diarrhea. When questioned about her past medical history, she said she had been to the emergency room on several occasions, but no problems were ever found.

"I think you have a condition called generalized anxiety disorder," the doctor said. "Here is a prescription for Ativan – it is very effective at controlling the symptoms of anxiety."

At her next visit a month later, the patient was feeling better and asked the doctor to renew her prescription. The doctor said he would do that, but first he wanted to test her urine to see that she was taking the drug as directed. He handed her a test requisition and told her to take it to the lab downstairs.

A laboratory report arrived 2 days later – the screening test for benzodiazepines was negative.

- Can you explain this negative test result?
- How would you resolve the situation?

Discussion

There are several possible explanations for the negative screening test. Ativan® is the proprietary name for lorazepam. When beginning therapy with lorazepam, the daily dose is in the range of 2–3 mg/d. This is a low dose, and the amount of drug excreted in urine may be below the detection level for some screening assays.

Lorazepam is conjugated to glucuronic acid before excretion. However, most screening assays will only detect the free drug. If the urine sample is not treated with glucuronidase before analysis, the screening test may be negative, even though the conjugated drug is present.

The patient's compliance with the doctor's orders must be considered as well. A negative screening result would be expected if the patient is not taking the medication. Sometimes a patient will take her medication intermittently, and the window period when it can be detected will pass.

A sensible next step is to consult with the laboratory that performed the testing. Ask about the sensitivity of their screening assay for lorazepam and how often they see false-negative results. It may be helpful to analyze the sample by a second method (i.e., GC-MS or LC-MS/MS) which is more sensitive and reliable. Finally, it is important to ask the patient if she has been taking Ativan as prescribed.

Takeaway messages

- Benzodiazepines are subject to false-negative screening results more often than other drug groups.
- When test results are unexpected, contact the laboratory and ask about the performance of their screening assays. Consider repeating the analysis with a second method based on a different measurement principle.

Case 8.2   Too Much Information

A 58-year-old man developed muscle spasms in his lower back following surgery for a herniated disc. The surgeon prescribed Valium® (diazepam), 10 mg at bedtime, to relieve the spasms and help him to sleep. Three months later at a follow-up visit, the patient asked to have the Valium prescription renewed.

The surgeon said, "You know, benzos are addictive. I would prefer to get you onto a different medication, if you even need one. Before making any changes, let's do a couple of tests."

A screening test for benzodiazepines was ordered. Later that week, he received a laboratory report that showed the following results.

Analysis by LC-MS

Nordiazepam	180 ng/mL
Temazepam	420 ng/mL
Oxazepam	230 ng/mL

The surgeon called the laboratory and asked why he had received such a detailed report. After remaining on hold for 10 minutes, the lab supervisor told him that the additional drug tests had been run in error. The patient's urine specimen had been accidentally placed in the confirmation rack after the screening test was positive.

"What am I supposed to do with these results?" the surgeon asked. "My patient has been taking Valium for the past 3 months to relieve muscle spasms in his back. Now I find out that he is taking several other drugs as well."

The surgeon (who was already suspicious) called his patient and asked him where he was getting the temazepam and oxazepam.

"Honest, doc, the only thing I'm taking is the Valium you gave me for those muscle cramps in my back. And it really helps me to sleep at night."

- What is the most likely explanation for the patient's test results?

Discussion
Diazepam is metabolized to nordiazepam, temazepam, and oxazepam, the three compounds that were detected on the drug confirmation assay. The patient's claim that he was taking only diazepam is therefore the simplest explanation for these results. While it is possible that he was taking two or even three different medications, it is unlikely.

Temazepam and oxazepam may be prescribed for insomnia or to treat anxiety or alcohol withdrawal (oxazepam). Because the metabolites of diazepam are drugs in their own right, healthcare practitioners may assume that the patient is taking these medications separately. Benzodiazepines are commonly misused, and their unexpected presence may arouse the suspicion of doctors, nurses, and other healthcare providers.

Takeaway messages

- To correctly interpret drug analyses, one needs to know the expected metabolites of the parent drug. In some situations, those metabolites are also drugs themselves.
- Clinical laboratories occasionally perform tests that were not requested by the physician. This may be due to illegibly written orders or an internal error by the laboratory.

Case 8.3   Empty Medicine Bottle

A 47-year-old woman was found unconscious in bed by her housemate at 9 a.m. She had been unhappy since a recent break-up with her boyfriend and had talked about "ending it all." There was no suicide note, but an empty medicine bottle with the label torn off was sitting on the night table. Her housemate suspected an overdose of sleeping pills and called emergency health services.

The patient was assessed by paramedics and transported to a local hospital. On arrival, her level of consciousness had not changed. Blood was drawn for a panel of tests including serum alcohol, with the following result:

Ethanol	50 mg/dL (11 mmol/L)

A catheterized urine specimen was obtained and sent for rapid screening. Results were as follows.

Immunoassay

Barbiturates	Negative
Benzodiazepines	Negative
Opiates	Negative

- Could this be a simple case of intoxication with ethanol?
- Do the screening results rule out a drug ingestion?
- What additional information would you request?

Discussion

According to her housemate, the patient may have tried to commit suicide by taking an overdose of sleeping pills. Ethanol is present in her blood, but the concentration is too low to produce unconsciousness. Benzodiazepines and

Z-drugs are commonly prescribed hypnotic agents – however, the screening test for benzodiazepines was negative. Barbiturates and opioids can also produce coma, but they were not detected in the urine sample.

Many other drugs could be responsible for this presentation. A list of the patient's medications would be very helpful – this may be available from her family doctor or a prescription database maintained by the local health system. An emergency room nurse searched the patient's medication record and discovered a prescription for zopiclone, 7.5 mg at bedtime, for insomnia. The patient picked up a 1-month supply of the drug last week at the pharmacy.

The urine sample that had screened negative was sent for a comprehensive drug analysis. The results came back 2 days later.

Analysis by LC-MS

Zopiclone	940 ng/mL

Takeaway messages

- Screening tests fail to detect a number of drugs that are commonly prescribed. In particular, screening tests for benzodiazepines will not detect Z-drugs.
- A comprehensive drug analysis can identify compounds that are missed by the usual screening assays. Results are useful in determining the cause of a patient's intoxication, but they are not available in time to guide acute management.
- Combined drug and alcohol ingestions are a common occurrence.

Case 8.4   Problems at School

KT, a 17-year-old high school senior, was found napping at her desk by her math teacher. Over the past month, she had stopped participating in class discussions and had missed two homework assignments. The teacher sent her to see the school nurse, who asked KT about her health, home life, and whether she was having any personal issues. KT said that she felt OK and everything was fine at home, she was just stressed about passing her exams.

The nurse referred her to a local clinic for adolescent health. KT filled out a form with questions about her physical and mental health, then spoke to a social worker about what was happening in her life. Her vital signs, height and weight were recorded, and she provided a urine sample for some screening tests. Before leaving, she was given an appointment with the clinic doctor for the following week.

At her next clinic visit, KT was examined by the doctor. "Physically, you appear to be in excellent health," she said. "But I am concerned about the results of your screening tests. There is evidence of cannabis use, and you had a drug called etizolam in your system. How long have you been smoking pot"?

"I smoke a joint 2 or 3 times a week, usually with friends after school. It's harmless, and it relaxes me."

"And where did you get the etizolam"?

"Etiza-what? I've never heard of that drug. A friend of mine has some pills that help me to calm down – they're called Xanax. I buy them for $5 when I am feeling anxious, and they make me feel really good."

- What is etizolam?
- How is it sold and distributed?
- Is etizolam detected in screening and confirmation tests for benzodiazepines?

Discussion

Etizolam is very similar in structure to the benzodiazepines (Figure 8.7). It has the same mechanism of action – binding to and activating the GABA receptor – and it produces the same antianxiety and hypnotic effects. Etizolam is 6–10 times more potent than diazepam. The drug is available by prescription in Japan, Italy, and India, but it is not an approved medication in North America.

Etizolam and so-called designer benzodiazepines (which are not approved for medical use anywhere) can be purchased from selected internet sites, where they may be labeled as research chemicals. Etizolam may also be diverted from countries where it is legally prescribed. Pills that contain etizolam have been sold as Xanax® or another prescription benzodiazepine. There is no assurance that pills and powders obtained from illicit sources actually contain the drug that was advertised.

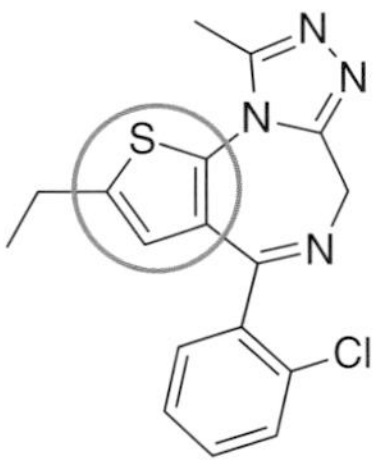

Etizolam

Figure 8.7 Structure of etizolam. A circle surrounds the thiophene ring that replaces the benzene ring in benzodiazepines – otherwise, the molecule is similar to alprazolam and triazolam (see Figure 8.3).

Screening tests for benzodiazepines will detect etizolam and most designer benzodiazepines due to similarities in structure. However, confirmation tests may be negative, because most laboratories do not include these drugs in their confirmatory assays.

In this case, KT was buying pills from her friend that were labeled as Xanax but contained etizolam instead. The laboratory that tested KT's urine sample was able to identify etizolam by LC-MS/MS.

Takeaway messages

- Etizolam is a benzodiazepine analog that is not approved for medical use in North America. Its physiological and psychological effects are similar to those produced by other benzodiazepines.
- The drug is obtained through illicit sources or by diversion. Illicit etizolam may be labeled as a familiar prescription benzodiazepine.
- A positive screening test followed by a negative confirmation test may indicate the presence of a designer benzodiazepine. Contact the toxicology laboratory to review which compounds they can identify.

Further Reading

Articles

Agarwal, S.D. and Landon, B.E. (2019). Patterns in outpatient benzodiazepine prescribing in the United States. *JAMA Netw. Open* 2 (1): e187399.

Carpenter, J.E., Murray, B.P., Dunkley, C. et al. (2019). Designer benzodiazepines: a report of exposures recorded in the National Poison Data System, 2014–2017. *Clin. Toxicol.* 57: 282–286.

Gunja, N. (2013). The clinical and forensic toxicology of Z-drugs. *J. Med. Toxicol.* 9: 155–162.

Book chapters

Hoffman, R.S., Nelson, L.S., and Howland, M.A. (2015). Benzodiazepines. In: *Goldfrank's Toxicologic Emergencies*, 10e (eds. R.S. Hoffman, M.A. Howland, N.A. Lewin, et al.), 1069–1075. New York: McGraw-Hill.

Lee, D.C. (2015). Sedative-hypnotics. In: *Goldfrank's Toxicologic Emergencies*, 10e (eds. R.S. Hoffman, M.A. Howland, N.A. Lewin, et al.), 1002–1012. New York: McGraw-Hill.

Websites

National Institute on Drug Abuse

Prescription CNS depressants
  www.drugabuse.gov/publications/drugfacts/prescription-cns-depressants

Centre for Addiction and Mental Health

Antianxiety medications (benzodiazepines)

www.camh.ca/en/health-info/mental-illness-and-addiction-index/anti-anxiety-medications-benzodiazepines

Videos

2-Minute Neuroscience – Benzodiazepines

www.neuroscientificallychallenged.com/blog/2-minute-neuroscience-benzodiazepines?rq=benzodiaze

Sedative Hypnotics: Benzodiazepines – CNS Pharmacology

www.youtube.com/watch?v=8c2EuTMcz_Q

Pharmacology – Benzodiazepines, Barbiturates, Hypnotics (Made Easy)

www.youtube.com/watch?v=4ZHudeMho8g

9

Other Sedative-Hypnotic Drugs

In addition to the benzodiazepines and Z-drugs, other sedative-hypnotic drugs are prescribed for therapeutic purposes. Many of these compounds have been replaced by newer drugs with better efficacy and safety profiles. All of them are subject to inappropriate use and, in the wrong hands, can be dangerous.

Barbiturates

The barbiturates are a family of drugs derived from barbituric acid. They all share the same 6-membered ring structure, but individual barbiturates differ in the side chains arising from one end of the molecule. The structures of several commonly used barbiturates appear in Figure 9.1.

Barbital, the first barbiturate with hypnotic properties, was developed in 1903. More than 50 barbiturates have been used for medical purposes, and about a dozen continue to be prescribed today.

How Barbiturates Work

Barbiturates depress the central nervous system (CNS) by enhancing the inhibitory effects of gamma-aminobutyric acid (GABA) and blocking the excitatory effects of glutamate.

Like benzodiazepines, barbiturates bind to the GABA-A receptor on the postsynaptic membrane of nerve cells. However, barbiturates bind to a different site and exert their effects in a different manner. When GABA binds to and activates the receptor, barbiturates increase the duration of chloride ion channel

An Introduction to Testing for Drugs of Abuse, First Edition. William E. Schreiber.
© 2022 John Wiley & Sons Ltd. Published 2022 by John Wiley & Sons Ltd.

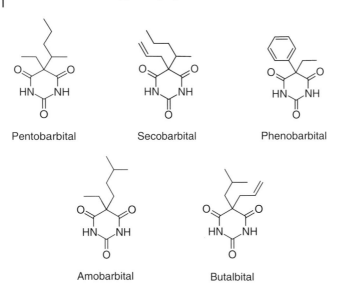

Figure 9.1 Structures of some common barbiturates. Each compound has two side chains attached to the ring that determine its unique structure.

opening. This causes hyperpolarization of the neuron and increases the threshold for generating an action potential. At high concentrations, barbiturates can open chloride channels without binding of GABA.

These drugs also interfere with transmission of nerve impulses mediated by the excitatory neurotransmitter glutamate. Barbiturates block several types of glutamate receptors that are located on the postsynaptic membrane. This prevents cations from flowing into the cell and initiating an action potential.

Physiological Effects

Barbiturates produce effects similar to those of benzodiazepines.

- Drowsiness
- Slurred speech
- Inability to concentrate
- Confusion
- Dizziness
- Loss of coordination
- Impaired memory
- Decreased blood pressure
- Slowed breathing

Therapeutic Uses

Since the introduction of benzodiazepines, which are much safer, barbiturates are not generally prescribed as sedatives or hypnotics. They continue to be used in the following situations.

- *Seizures* – phenobarbital is widely prescribed for the treatment of epilepsy. Pentobarbital is sometimes administered for emergency control of acute convulsive episodes, such as status epilepticus and eclampsia.
- *Anesthesia* – thiopental is a commonly used anesthetic agent.
- *Headache* – a preparation containing butalbital, acetaminophen, and caffeine is used to treat tension headaches.

Potential for Abuse

Barbiturates are less available to outpatients than they once were and are therefore not misused to the same extent as other drug groups. They are potent CNS depressants that, when taken in excess, can be lethal either by themselves or in combination with other drugs.

Metabolism

Barbiturates are converted to a number of metabolites. The side chains are oxidized to form alcohols and carboxylic acids, many of which are conjugated to glucuronic acid. Only small amounts of the parent compounds (typically <5%) are excreted unchanged. Phenobarbital is an exception – 25% of the parent drug appears in urine. Table 9.1 gives the elimination half-lives for common barbiturates.

Table 9.1 Elimination half-life data for common barbiturates.

Drug	Half-life (h)
Amobarbital	15–40
Butabarbital	34–42
Butalbital	35–88
Pentobarbital	15–48
Phenobarbital	48–144
Secobarbital	22–29
Thiopental	6–46

Source: Baselt RC. Disposition of Toxic Drugs and Chemicals in Man, 9th edn. Seal Beach, CA: Biomedical Publications, 2011.

Table 9.2 Cross-reactivity of selected barbiturates in a commercial immunoassay.

Barbiturate	% Cross-reactivity
Amobarbital	57
Butabarbital	73
Butalbital	66
Pentobarbital	79
Phenobarbital[a]	21–39
Secobarbital	100
Thiopental	0.7

Cross-reactivity was determined at the 200 ng/mL cut-off for a positive test.
[a] Observed range.
Source: Siemens ADVIA Chemistry Systems, Barbiturate_2 (BARB_2).

Testing for Barbiturates

Screening
Screening immunoassays are designed to detect the parent drugs in urine (Table 9.2). Although barbiturates are transformed to other metabolites, they are taken in high enough doses that the unchanged drug can be identified. The cut-off for a positive result is 200 or 300 ng/mL.

Confirmation
Many different assays have been developed for barbiturates. Gas chromatography is usually combined with a flame ionization detector or mass spectrometer. Liquid chromatography-based assays employ an ultraviolet detector or mass spectrometer to measure the amount of each drug that is present.

Window of Detection
Short-acting barbiturates, such as pentobarbital and secobarbital, can be detected up to 4–6 days after use. Phenobarbital, which is long-acting, may be detected from 10 to 30 days following use.

Interferences

False-positive results in the screening immunoassay have been reported with ibuprofen and naproxen, both of which are nonsteroidal antiinflammatory drugs.

Gamma-Hydroxybutyric Acid (GHB)

GHB occurs naturally within the human body, albeit at low concentrations. When consumed in amounts of 1 g or more, it acts as a CNS depressant. Its structure resembles that of GABA, the major inhibitory neurotransmitter (Figure 9.2). Two similar compounds, gamma-butyrolactone and 1,4-butanediol, are enzymatically converted to GHB in vivo and produce the same effects.

Research into GHB as a human drug was first reported in the 1960s. Since that time, it has been used as an anesthetic, a muscle-building supplement, and a treatment for alcohol withdrawal. In the 1990s, GHB became popular with young adults for its euphoric effects.

How GHB Works

The effects of GHB are mediated by two different receptors to which it binds. The GABA-B receptor is normally activated by the inhibitory neurotransmitter GABA. It is a G-coupled protein that controls the activity of potassium channels in the cell membranes of neurons. When activated, the receptor causes these channels to open, allowing potassium ions to flow out of the cell. The change in electrical potential exerts an inhibitory effect on transmission of nerve impulses. GHB has a weak affinity for the GABA-B receptor, and at high concentrations it mimics the effects of GABA.

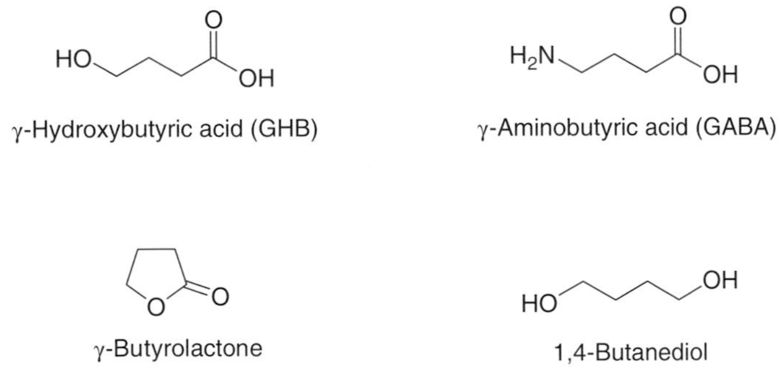

γ-Hydroxybutyric acid (GHB)

γ-Aminobutyric acid (GABA)

γ-Butyrolactone

1,4-Butanediol

Figure 9.2 Structures of gamma-hydroxybutyric acid (GHB) and the inhibitory neurotransmitter gamma-aminobutyric acid (GABA). Gamma-butyrolactone and 1,4-butanediol are industrial chemicals that are metabolized to GHB in vivo.

A more specific GHB receptor is widely distributed throughout the brain. Activation of this receptor stimulates release of dopamine, producing a euphoric state.

Physiological Effects

GHB produces the following effects.

- Drowsiness
- Relaxation
- Sense of well-being
- Loss of inhibitions
- Dizziness
- Amnesia
- Loss of coordination
- Nausea
- Hypothermia
- Decreased blood pressure and heart rate
- Slowed breathing
- Loss of consciousness

Therapeutic Uses

GHB is used as a treatment for narcolepsy but is only available to patients enrolled in a restricted access program. When consumed for any other reason (i.e., recreational purposes), it is considered a Schedule I drug.

Potential for Abuse

GHB comes in two forms: an odorless, colorless liquid and a white powder. The drug is most often used by teenagers and young adults attending bars, nightclubs, and parties. When mixed with alcoholic beverages, its presence is difficult to detect. The effects of GHB begin within 10–20 minutes following ingestion and typically last for 1–4 hours.

GHB has been used to incapacitate victims prior to sexual assault, earning it a place among so-called "date rape" drugs. After the effects wear off, victims of sexual assault may not remember the event. The rapid elimination of the drug leaves only a short time window in which to obtain blood or urine samples for analysis. Very high doses are dangerous and may be fatal, especially when intoxicants such as alcohol or other CNS depressants are also consumed.

Metabolism

GHB is metabolized to succinic acid and enters the tricarboxylic acid cycle, where it serves as an energy substrate. Less than 5% of a dose is excreted unchanged in urine. The elimination half-life is 30–60 minutes.

Testing for GHB

GHB is not detected in routine drug screens, so analysis for the drug must be specifically requested. Both gas chromatography and liquid chromatography-based methods can identify GHB and measure its concentration. Blood and urine samples should be collected as soon as possible following a suspected ingestion, and at the latest within 8 hours and 12 hours, respectively.

Since GHB is produced by the body, evidence of its use requires a concentration above normal endogenous levels. Recommended cut-off values for a positive result are >5 µg/mL in serum and >10 µg/mL in urine samples.

Chloral Hydrate

Chloral hydrate was among the first drugs with sedative-hypnotic properties to be used in medical practice. It was synthesized in 1832 and became popular as a treatment for anxiety and insomnia in the latter half of the nineteenth century. It remains available by prescription, although its uses are limited.

Following ingestion, the drug is metabolized within minutes to trichloroethanol, which acts as a CNS depressant (Figure 9.3). The effects of ethanol and chloral hydrate are synergistic, since each drug interferes with the metabolism of the other. Addition of chloral hydrate to ethanol-containing beverages (known as a Mickey Finn) can cause the drinker to lose consciousness.

The half-life of trichloroethanol is 6–10 hours. Most of this compound is further metabolized to trichloroacetic acid or conjugated to glucuronic acid prior to excretion. Testing for trichloroethanol in blood or urine is available through specialized toxicology laboratories.

Chloral hydrate   Trichloroethanol

Figure 9.3 Structures of chloral hydrate and its active metabolite trichloroethanol.

Case Studies

Case 9.1   Frat Party

BR, a 19-year-old college sophomore, attended a social event at a fraternity house on the edge of campus. Beer was flowing from a keg, and mixed drinks were being served at the adjoining bar. The social director invited her to see the house clubroom and said he would refresh her drink. As the evening progressed, she began to feel sleepy, then very drowsy before losing consciousness. She woke up on a bed in a room she did not recognize, naked from the waist down.

When BR got home, she called her best friend who said, "Oh no! You may have been raped." The best friend came by, picked up BR and drove her to the student health center. The doctor on duty asked her what happened.

"I was at the party, having a good time and talking to this guy. Suddenly I felt sleepy, and I guess I just blacked out. The next thing I remember is waking up – my pants were off, and I didn't know where I was."

"Believe it or not," the doctor said, "you are the second patient I've seen who attended that party – similar story. Let's bring our nurse in to help with the exam."

- Could this be a case of drug-facilitated sexual assault?
- If so, what samples would you collect? What drugs would you test for?

Discussion

It is possible that sexual activity occurred while BR was unconscious. Because she was at a party where alcohol was being served, it is important to know if she drank heavily, to the point of passing out. In fact, she claimed to have consumed one beer and one mixed drink over a 2-hour period, which is not enough to produce unconsciousness. Someone may have spiked her drinks with a drug in order to reduce her inhibitions or cause her to fall asleep.

Sexual assault is a crime, whether the victim is awake or asleep, so any evidence that is gathered may be presented in court. Both urine and blood samples should be collected. Drugs may be detectable in blood up to 24 hours following ingestion. Urine samples may test positive up to 5 days after a drug was taken.

Three compounds are frequently referred to as date rape drugs: GHB, flunitrazepam, and ketamine. Some laboratories offer this specific panel in their test menu. Other drugs have been implicated in drug-facilitated sexual assault as well, including:

- benzodiazepines
- antidepressants

- hallucinogens
- opioids
- over-the-counter sleep aids and antihistamines.

Screening assays for benzodiazepines, opioids, cannabinoids, and amphetamines (for MDMA) in urine will detect many of these substances. Flunitrazepam is a benzodiazepine, but it may not be detected in the screening assay because the usual dose is in the 1–2 mg range. Testing for GHB and ketamine requires analysis by GC-MS or LC-MS/MS. Unless the laboratory runs a comprehensive drug screen, a number of these substances will be missed.

In this case, testing of the patient's urine revealed the following.

Immunoassay

Amphetamines	Negative
Barbiturates	Negative
Benzodiazepines	Negative
Cannabinoids	Negative
Opiates	Negative

Analysis by LC-MS

Gamma-hydroxybutyric acid (GHB)	450 µg/mL

A police investigation followed, and criminal charges were laid against two members of the fraternity.

Takeaway messages
- Drug-facilitated sexual assault often takes place in settings where alcohol is consumed.
- Drugs that are used for this purpose typically cause decreased inhibitions, loss of muscle control, drowsiness/unconsciousness, and retrograde amnesia.
- Standard screening tests will not detect GHB or other drugs that are frequently implicated. Order individual tests by name.
- Sample collection, transport, storage, and analysis should follow chain of custody rules, because test results may be presented as evidence in court proceedings.

Case 9.2   Sleepy Kid

PJ, a 15-year-old boy, was found sleeping on the living room couch by his mother at 3 pm. He is very active and usually spends his afternoons outside riding his bike or hanging with school friends. When she tried to wake him, he muttered something unintelligible and rolled over without opening his eyes. His mother became concerned that he was sick and called the doctor's office. They advised her to take PJ to the emergency room.

Her husband hauled PJ into the car, and the three of them drove to the local urgent care center. After taking a brief history from his parents and doing a cursory examination, the physician said that he suspected PJ may be intoxicated.

"Does he drink, smoke grass or take any illegal drugs?" the doctor asked.

"Not that we know of" his mother answered.

"Any prescription medications lying around the house?"

"No, not really . . . but his younger brother is epileptic and takes Luminal to prevent seizures."

The doctor said, "OK, I'm going to order tests for several drugs, including Luminal, to see if that's the cause of his drowsiness."

- What is Luminal®?
- Why would someone take this drug?
- How would you test for its presence?

Discussion

Luminal is the proprietary name for phenobarbital. It is most commonly pre-scribed for the management of several types of seizures. Phenobarbital has also been used to treat anxiety and insomnia, but benzodiazepines and Z-drugs are safer and have replaced barbiturates for these indications.

People misuse barbiturates because they produce effects similar to those of alcohol – happiness, relaxation, and reduced inhibitions. The drug can be diverted and sold illegally, or it can be obtained from a family member who has a prescription.

In this case, PJ's younger brother has a supply of phenobarbital to treat epilepsy. One of his friends told him that "barbs will get you really high." PJ has been stealing his brother's pills and taking them for the past week.

Barbiturates can be detected on a urine drug screen. Quantitative measurements of serum phenobarbital are also available in many laboratories, because the drug is routinely monitored as an antiepileptic medication. PJ's blood test results were as follows.

Serum toxicology

Phenobarbital	48 µg/mL (210 µmol/L)
Ethanol	Not detected

The therapeutic range for phenobarbital in serum is 15–40 µg/mL (65–172 µmol/L), so this level would be considered toxic

Takeaway messages
- When investigating a suspected drug ingestion, ask about medications taken by family members and close associates.
- Rapid blood tests are not available for most drugs of abuse. Phenobarbital is an exception – some clinical laboratories can perform the analysis and report results within 1–2 hours.

Case 9.3   Celebrity

In the predawn hours of August 5, 1962, Marilyn Monroe was discovered deceased in the bedroom of her Los Angeles home. A collection of sedatives, stimulants, and sleeping medications was found on her bedside table. She was among the most popular movie stars in the world in the 1950s and early 1960s and the object of public fascination. During the last few years of her life, she suffered from a variety of mental health problems as well as substance abuse.

According to the postmortem toxicology report, the following drugs were present in her blood.

Pentobarbital	45 mg/L
Chloral hydrate	80 mg/L

The autopsy report concluded that the cause of death was acute barbiturate poisoning.

- What is your assessment?

Discussion
Concentrations of pentobarbital in excess of 15–25 mg/L are considered potentially lethal. A chloral hydrate level (measured as trichloroethanol) of 60–100 mg/L or higher may be fatal as well. Acute poisoning from a mixed drug overdose might be a better description of the cause of death.

While there is general agreement that Ms Monroe died from an overdose of drugs, the actual manner of her death has been a source of controversy.

According to the autopsy report, the mode of death was "probable suicide." However, some authors maintain that she was murdered, while others claim that the death was accidental and made to look like a suicide. None of these conspiracy theories has been proven.

Takeaway messages
- Pentobarbital is a potent drug. When taken in large doses or in combination with other CNS depressants, it can be deadly.
- Celebrities receive a great deal of attention. This can affect their perception of reality and exacerbate mental, emotional, and behavioral problems.
- People love gossip, whether it is true or not.

Further Reading

Articles

Busardo, F.P. and Jones, A.W. (2019). Interpreting gamma-hydroxybutyrate concentrations for clinical and forensic purposes. *Clin. Toxicol.* 57: 149–163.
Lopez-Munoz, F., Ucha-Adabe, R., and Alamo, C. (2005). The history of barbiturates a century after their clinical introduction. *Neuropsychiatr. Dis. Treat.* 1: 329–343.

Book chapters

Farmer, B.M. (2015). γ-Hydroxybutyric acid. In: *Goldfrank's Toxicologic Emergencies*, 10e (eds. R.S. Hoffman, M.A. Howland, N.A. Lewin, et al.), 1124–1128. New York: McGraw-Hill.
Lee, D.C. (2015). Sedative-hypnotics. In: *Goldfrank's Toxicologic Emergencies*, 10e (eds. R.S. Hoffman, M.A. Howland, N.A. Lewin, et al.), 1002–1012. New York: McGraw-Hill.

Websites

National Institute on Drug Abuse
Prescription CNS Depressants
   www.drugabuse.gov/publications/drugfacts/prescription-cns-depressants

Centre for Addiction and Mental Health
   GHB www.camh.ca/en/health-info/mental-illness-and-addiction-index/GHB

Videos

2-Minute Neuroscience – GABA
   www.youtube.com/watch?v=bQIU2KDtHTI
Pharmacology – Benzodiazepines, Barbiturates, Hypnotics (Made Easy)
   www.youtube.com/watch?v=4ZHudeMho8g

10

Opioids

The opioids are a diverse group of compounds that have in common the ability to bind and activate opioid receptors. In nature, opioids are derived from the opium poppy, *Papaver somniferum* (Figure 10.1). The poppy heads contain a white milky substance called latex that is harvested and dried to produce opium. Opium is rich in alkaloids – about 10% of its weight consists of morphine. Lesser amounts of codeine and thebaine (a precursor of semisynthetic opioids) are present. The term "opiates" is used to describe these naturally occurring compounds.

Minor variations in structure produce a series of related opioids with differing biological activities. Hydromorphone, hydrocodone, oxymorphone, oxycodone, and heroin are all semisynthetic opioids built on the alkaloid structure of morphine (Figure 10.2).

Years of pharmaceutical research have produced numerous synthetic compounds with opioid-like effects. A number of these have been introduced into clinical practice. The most widely used synthetic opioids are methadone, fentanyl, tramadol, meperidine, and propoxyphene (Figure 10.3).

How Opioids Work

At the molecular level, opioids bind to specific opioid receptors located in the brain, spinal cord, peripheral nerves, and intestinal tract. There are three well-established opioid receptors, named mu, kappa and delta, of which mu is considered the most important. When an opioid ligand binds to its receptor on the cell surface, the signal is transmitted into the cell through coupling of the receptor to G proteins.

Opioids exert their effects by controlling the movement of calcium and potassium ions across the cell membrane. Calcium influx is required to trigger the

An Introduction to Testing for Drugs of Abuse, First Edition. William E. Schreiber.
© 2022 John Wiley & Sons Ltd. Published 2022 by John Wiley & Sons Ltd.

Figure 10.1 Picture of the opium poppy, *Papaver somniferum. Source:* Radovan 1/Shutterstock.

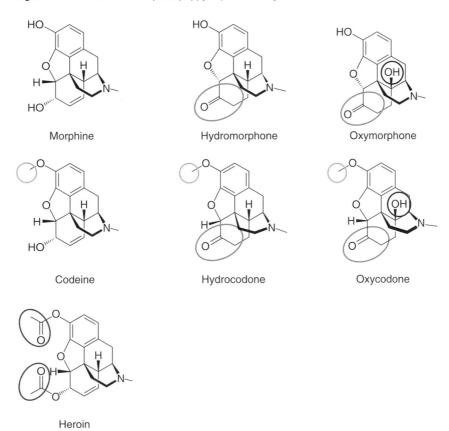

Figure 10.2 Structures of morphine, codeine, and commonly used semisynthetic opioids. Circles and ellipses show the differences between morphine and each opioid.

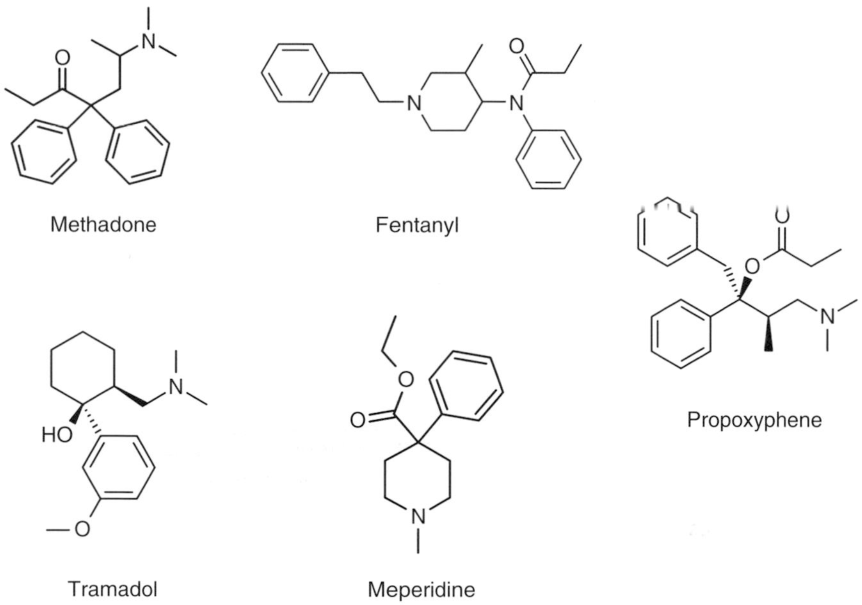

Figure 10.3   Structures of commonly prescribed synthetic opioids. None of these compounds has the characteristic interlocking ring structure of morphine.

release of neurotransmitters at presynaptic nerve endings. Activated opioid receptors block voltage-gated calcium channels and prevent calcium from entering the nerve cell. This inhibits neurotransmitter release, thereby reducing transmission of signals from one neuron to the next. Opioids also stimulate ion channels that pump potassium out of the cell. This creates hyperpolarization of the neuronal membrane, which interferes with propagation of the action potential.

The affinity of individual opioids for the opioid receptors determines their biological effect. Opioids are classified as full agonists (e.g., morphine) or partial agonists (e.g., buprenorphine) based on how strongly they bind to and activate the opioid receptor. Some opioids are classified as antagonists, because they can displace an agonist and reverse its activity (e.g., naloxone – see Figure 10.4).

The relative potency of common opioids is given in Table 10.1.

Physiological Effects

The major effects of opioids on the body are:

- analgesia
- euphoria

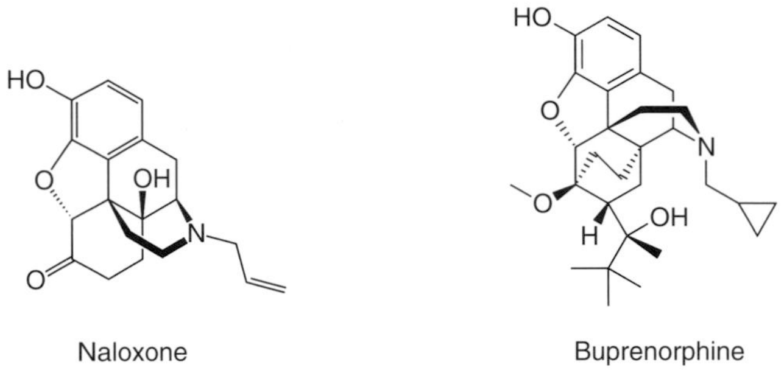

Naloxone   Buprenorphine

Figure 10.4 Structures of naloxone, an opioid antagonist, and buprenorphine, a partial agonist/antagonist.

Table 10.1 Relative potency of opioids.

Opioid	Equivalent oral dose[a]	Relative potency
Codeine	200 mg	0.15
Morphine	30 mg	1
Hydrocodone	30 mg	1
Oxycodone	20 mg	1.5
Oxymorphone	10 mg	3
Hydromorphone	6 mg	5
Heroin	6–15 mg	2–5
Fentanyl	0.3–0.6 mg	50–100

[a] Equivalent doses for heroin and fentanyl are based on intravenous and transdermal administration, respectively.
Source: Data are from multiple sources.

- sedation, coma
- nausea and vomiting
- constipation
- suppression of coughing
- constriction of pupils
- respiratory depression.

Therapeutic Uses

- *Pain management* – opioids are the strongest analgesic medications available. There are numerous choices of drug and formulation that physicians can prescribe for their patients, both in hospital and as outpatients.

- *Medical procedures* – opioids may be used as part of anesthesia during surgery and other procedures, such as colonoscopy.
- *Antidiarrheal* – gastrointestinal motility is reduced by opioids, making them effective in the treatment of diarrhea.
- *Cough suppressant* – opioids suppress the cough reflex and are found in a number of cough medicines.
- *Antidote to acute opioid intoxication* – naloxone is a full antagonist that is administered in cases of suspected overdose with an opioid.

Nearly 50 million Americans (15.0% of the population) filled at least one prescription for an opioid in 2018. The average number of prescriptions per patient was 3.4.

Potential for Abuse

Opioids have a high potential for misuse and addiction. They are powerful analgesics and produce a pleasurable sense of well-being and euphoria. Tolerance to these effects develops over several weeks of regular use, requiring stronger doses to achieve the same subjective sense of relief. Both psychological and physical dependence on opioids is common. This can interfere with cessation of therapy, and it can lead to long-term addiction to both prescription and illicit drugs of this class.

In 2018, an estimated 9.9 million people in the United States aged 12 years and older (3.6% of this population) reported misuse of prescription opioids in the past year. Misuse was slightly higher among males than females, and the highest rates of misuse were in people aged 30–34 years (5.6%) and 18–25 years (5.5%). Another 0.8 million people aged 12 years and older (0.3% of this population) reported heroin use in the past year.

Opioid use has a major impact on the US health-care system. Drug-related poisonings accounted for more than 577 000 emergency department visits in 2016. One-third of these visits were due to opioid consumption. In 2017, a total of 47 600 people died from drug overdoses involving opioids. This figure represents 68% of all drug overdose deaths in the United States for that year.

Individual Drugs

There are numerous compounds within the opioid class – an encyclopedic listing can be found in other sources. Here we highlight the drugs that are likely to be encountered when testing human samples.

Naturally Occurring Opioids

Morphine

Morphine is the opioid to which other drugs of this class are most frequently compared. Its structure of five interlocking rings is found in naturally occurring and semisynthetic opioids. The drug is formulated in tablets for oral use and in solution for intravenous or intramuscular injection.

Morphine is prescribed for moderate to severe pain. It has a half-life of 1.3–6.7 hours and is metabolized by conjugation to glucuronic acid, producing morphine-3-glucuronide (inactive) and morphine-6-glucuronide (active). Small amounts of hydromorphone (up to 1–2% of morphine metabolites) may be produced in patients on long-term morphine treatment.

Codeine

Codeine is prescribed for the relief of mild to moderate pain. It is widely available in formulations with acetaminophen or aspirin as an analgesic. It is also included in cough or cold medications as a cough suppressant.

Codeine is metabolized to morphine, which is about 10 times more potent, so it is often considered a prodrug. The rate at which it is converted depends on the activity of the enzyme CYP2D6. This enzyme has multiple alleles which vary in activity. In patients who are rapid metabolizers, the rate of morphine production is high, which may lead to toxic effects. In slow metabolizers, there may be little conversion of codeine to morphine and a corresponding lack of effectiveness.

The half-life of codeine is 1.2–3.9 hours. The major metabolites in urine are codeine, norcodeine, and morphine conjugates. A day after ingestion, the morphine concentration in urine may exceed that of codeine, and codeine may vanish altogether after 2 days. Trace amounts of hydrocodone can also be produced.

Semisynthetic Opioids

Heroin

Heroin (diacetylmorphine) is a semisynthetic opioid with 2–5 times the potency of morphine. Because the compound is less polar, it passes through the blood–brain barrier and into nervous tissue more readily than other opioids. Heroin is an illicit drug and is typically consumed by intravenous injection or nasal insufflation ("snorting").

Conversion to 6-acetylmorphine is very rapid, with a half-life of 2–6 minutes, and subsequent metabolism to morphine has a half-life of 6–25 minutes. Analysis of urine from a heroin user detects mainly morphine and its conjugates. The presence of 6-acetylmorphine is unequivocal proof of heroin use, but this compound disappears from urine within the first day following consumption.

Hydrocodone and Oxycodone

These two semisynthetic opioids are of similar and slightly greater potency, respectively, than morphine when taken orally. Both drugs are prescribed for relief of moderate to severe pain. In 2017, hydrocodone was the most commonly prescribed drug in the United States. Half-lives are 3.4–8.8 hours for hydrocodone and 3–6 hours for oxycodone.

Hydrocodone is metabolized in vivo to a more active opioid, hydromorphone, which is conjugated and excreted in urine. Similarly, oxycodone is converted to oxymorphone, and conjugates of both drugs are excreted. Norhydrocodone and noroxycodone are other metabolites that may appear.

Hydromorphone and Oxymorphone

Hydromorphone and oxymorphone are indicated in the treatment of moderate to severe pain. They have longer half-lives than morphine (3–9 hours and 4–12 hours, respectively) and provide greater pain relief.

Hydromorphone is converted to 6-hydromorphol, and oxymorphone is metabolized to 6-oxymorphol. Both metabolites and their parent compounds are conjugated and excreted in urine.

Synthetic Opioids

Methadone

Methadone's major use is in the treatment of patients with an opioid addiction. It does not produce the same sense of euphoria as other opioids, but it does prevent withdrawal symptoms. It is also prescribed for pain management. The half-life of methadone, 15–55 hours, is longer than that of other opioids.

The main metabolite of methadone is 2-ethylidene-1,5-dimethyl-3,3-diphenyl-pyrrolidine (EDDP). Screening immunoassays for methadone measure this compound. The presence of EDDP in urine demonstrates that methadone was taken and metabolized within the body. If a urine sample contains only methadone, the drug was added after sample collection, likely in an attempt to show compliance.

Fentanyl

Fentanyl is a versatile drug that is used in many medical applications. It is administered intravenously for surgical anesthesia and outpatient procedures, orally as a lozenge for relief of breakthrough pain, and transdermally as a patch on the skin for long-term pain control. Fentanyl has at least 50 times the potency of morphine and a short duration of action. Its half-life is 3–12 hours. Numerous analogs of fentanyl have been synthesized, and some of these are used in medical and veterinary practice.

In the past decade, increasing amounts of fentanyl have entered the illicit drug market in North America. Users may or may not be aware that they are taking fentanyl, as it is often mixed with heroin, cocaine, and other street drugs. Because it is such a potent drug, thousands of overdose deaths each year are now attributed to fentanyl.

The major metabolite is norfentanyl. Immunoassays specific for fentanyl or its metabolite are available. Identification of fentanyl analogs requires GC-MS or LC-MS/MS analysis.

Other Synthetic Opioids

Meperidine, which is less potent than morphine, has a half-life of 2–5 hours. It is given both intravenously and orally. The drug is metabolized to normeperidine, which also has analgesic effects. Both the parent drug and metabolite are excreted in urine.

Tramadol has a potency similar to codeine. O-desmethyltramadol, its main metabolite, exhibits a greater analgesic effect than the parent drug. Tramadol is considered safer than other opioid narcotics and has less potential for abuse.

Propoxyphene is another mild analgesic, weaker than codeine. It is converted to the active metabolite norpropoxyphene. The drug is less frequently prescribed than in the past, but misuse remains a problem.

Of these three opioids, only propoxyphene is routinely tested by a screening immunoassay.

Opioid Antagonists

Naloxone

Naloxone is as an opioid receptor antagonist. It is given as an antidote for acute opioid toxicity and can reverse the signs of toxicity within seconds to minutes. Naloxone is usually administered by intravenous, intramuscular or intranasal routes.

The drug is similar in structure to oxymorphone, and it works by displacing opioids from the receptor. A side-effect of treatment is that naloxone can trigger withdrawal symptoms in patients who are opioid dependent. It is not a drug of abuse, and testing for naloxone is rarely needed.

Buprenorphine

Buprenorphine is a partial agonist of the mu receptor and an antagonist of the kappa receptor. Like methadone, it is used to treat patients addicted to opioids. The main metabolite is norbuprenorphine, which also has biological activity. A screening assay that detects buprenorphine, norbuprenorphine, and their glucuronide conjugates is available.

Testing for Opioids

Screening

Screening tests for opioids are usually performed by immunoassays with commercial reagents. The cut-off for a positive result is 300 ng/mL in clinical samples. Workplace drug testing has a much higher cut-off of 2000 ng/mL, which is set by the Substance Abuse and Mental Health Services Administration (SAMHSA) guidelines. This higher value is used to avoid false-positive results in people who have consumed poppy seeds.

Tests are configured to detect morphine, codeine, and heroin/6-acetylmorphine at or near the defined cut-off concentration. Hydrocodone and hydromorphone are also detected in opioid screening assays, but with lower sensitivity. Oxycodone and oxymorphone react weakly in screening tests, and synthetic opioids are not detected at all. Table 10.2 shows the cross-reactivity of one commercial assay with different opioids.

To overcome these limitations, immunoassays for opioids that are missed by the standard screening test have been developed. Assays specific for oxycodone/oxymorphone, hydrocodone/hydromorphone, methadone, fentanyl, buprenorphine, and other opioids are available. Health-care providers can consult with their clinical laboratory to review which assays are performed.

Table 10.2 Cross-reactivity of opioids in a commercial immunoassay.

Opioid	% Cross-reactivity
Morphine	100
Codeine	125
6-Acetylmorphine	81
Heroin	53
Dihydrocodeine	50
Hydrocodone	48
Hydromorphone	57
Morphine-3-glucuronide	81
Morphine-6-glucuronide	47
Oxycodone	3.1
Oxymorphone	1.9

Source: CEDIA Opiate Assay, Microgenics Corporation.

Confirmation

A positive immunoassay should not be taken as definitive proof that an opioid is present. Confirmation testing is usually performed by gas chromatography-mass spectrometry (GC-MS) or liquid chromatography-mass spectrometry (LC-MS). These techniques can identify individual opioids and their metabolites, and they can measure their concentrations at values below the cut-off for screening tests.

Most GC-MS and LC-MS assays are developed in-house by laboratory staff. The test menu typically includes common opioids such as morphine, codeine, and 6-acetylmorphine, but coverage of semisynthetic and synthetic opioids varies from one clinical laboratory to the next.

Window of Detection

As a rule, opioids can be detected in urine samples up to 3 days after drug ingestion. Exceptions to this rule are 6-acetylmorphine, which disappears within a day of heroin use, and codeine, which may be undetectable after 2 days. Methadone can be detected up to 7 days after use in some patients.

Interferences

Consumption of poppy seeds can give a positive urine drug test for opioids, because they contain small amounts of morphine and codeine. False-positive results can occur with screening tests due to other drugs in a urine sample. The paper by Saitman and colleagues (J Anal Toxicol 2014;38:387–396) provides a detailed list of interferences that have been reported.

Case Studies

Case 10.1   Bad Coke

At 3 p.m., emergency health services was notified about a possible drug overdose in an alleyway behind the train station. Minutes later, paramedics arrived to find two men slumped against a wall, neither showing any signs of life. Syringes, needles, and other drug paraphernalia were scattered on the pavement nearby. A dose of naloxone was injected into the upper arm of each man. One of them gasped and began to move; the other remained motionless. A second dose of naloxone was given with no effect and, following phone consultation with an emergency physician, the man was pronounced dead.

The first man was taken to a local hospital emergency room. When questioned, he said that he and his partner had "scored some coke" and decided to "mainline it." A urine drug screen was ordered and returned the following results.

Immunoassay

Amphetamines	Negative
Benzodiazepines	Negative
Cocaine	Positive
Opiates	Negative

The second man was taken to the morgue. Femoral vein blood was drawn for toxicology studies, which showed the following.

Analysis by LC-MS

Cocaine	0.1 mg/L
Benzoylecgonine	0.8 mg/L
Fentanyl	0.012 mg/L

- How do you explain a negative screening result for opiates in the survivor when fentanyl was present in the deceased?
- Which drug is the more likely cause of death?
- Why is fentanyl present in a powder sold as cocaine?
- Is it wise to give naloxone before confirming that someone has taken opioids?

Discussion

The screening immunoassay for opiates is designed to detect morphine, codeine, and 6-acetylmorphine. Although fentanyl is an opioid, its structure is different from naturally occurring opiates, so it does not react in the opiates assay. It was identified in postmortem blood by LC-MS, which can detect a number of different opioids.

Blood concentrations of fentanyl up to 0.002 mg/L are considered therapeutic. Values above that may be toxic or lethal, and fentanyl concentrations of 0.02 mg/L or higher in recreational users (i.e., without medical support) are nearly always fatal. A concentration of 0.012 mg/L is very high, and it is reasonable to ascribe fentanyl as the cause of death. Cocaine may be fatal at blood levels as low as 0.5–1.0 mg/L of benzoylecgonine, but values are typically higher.

Fentanyl is sometimes mixed with other illicit drugs before being sold on the street. Many fentanyl overdoses have occurred in users who thought they were taking cocaine, amphetamines, heroin or another drug. This practice has been responsible for numerous fatalities.

People who overdose on opioids and lose consciousness may have only minutes before they stop breathing and die. There is no time to lose – rapid treatment with naloxone is their best chance of survival. Any side-effects of naloxone use can be medically managed after the patient is out of danger.

Takeaway messages

- Fentanyl is an extremely potent opioid. It is responsible for thousands of overdose deaths in North America every year.
- Fentanyl is not detected by screening tests for opiates. Immunoassays that are specific for fentanyl or its metabolite norfentanyl are available in some clinical laboratories.
- Rapid treatment with naloxone can reverse the effects of fentanyl and other opioids and can save a person's life.

Case 10.2 Bagels for Lunch

DM is a 39-year-old man with a long history of buying and injecting street drugs. Four weeks ago, he entered a residential treatment program for sub-stance abusers after overdosing on heroin. On Saturday, he went out on a day pass. When he returned for dinner, one of the staff members thought that he was high and asked him to give a urine sample for testing. The test strip for opiates turned positive.

"D, did you visit your old friends at the shooting gallery today?" the staff member asked.

"No, man, I've given that up – going straight now."

"Well how do you explain this?" He showed D the positive test strip.

"Oh, that's nothing. I went to the deli and ate some bagels for lunch – you know, the kind with poppy seeds. They're my favorites."

- Can eating poppy seeds cause a positive screening test for opiates? Why?

The staff member sent D's urine sample to a toxicology lab for further test-ing. The results came back at the end of the week.

Analysis by LC-MS

Morphine	3200 ng/mL
Codeine	270 ng/mL
6-Acetylmorphine	45 ng/mL

- What do these test results indicate?

Discussion

Poppy seeds come from the same plant that produces opium alkaloids. During harvesting, poppy seeds can absorb or become coated by the plant's opium-containing residue. Most of this residue is washed away when the seeds are processed, but small amounts of morphine and codeine remain.

Poppy seeds are found in a number of baked foods – breads, muffins, bagels, pastries, and cakes – as well as salad dressings. After consuming these foods, urine drug tests for opiates may turn positive. It depends upon the concentration of opium alkaloids in the seeds, the number of seeds that are ingested, and the length of time between eating the seeds and collecting the sample.

The detection limit in workplace screening tests for opiates is 2000 ng/mL, which is much higher than the threshold for other drugs of abuse (see Table 1.2 in Chapter 1). This level was chosen to minimize positive results due to poppy seed consumption. A cut-off of 300 ng/mL is more often used in testing for clinical purposes, including substance abuse programs. Eating poppy seeds could be the cause of a positive test at this lower cut-off.

Unfortunately for DM, the confirmation test by LC-MS does not support his poppy seed theory. The presence of 6-acetylmorphine is proof that he took heroin. If the urine sample had been collected on the following day, 6-acetylmorphine would be undetectable and the levels of morphine and codeine would be lower, suggesting that the poppy seed bagels were responsible for his positive opiates test.

Takeaway messages

- Poppy seeds can cause a positive test for opiates in urine. To prevent this from happening, avoid poppy seed-containing foods for 2–3 days prior to sample collection.
- Heroin use and poppy seed consumption can produce similar results for morphine and codeine in urine samples. Some drug users may offer the "poppy seed defense" to explain positive test results.

Case 10.3 A Fruity Urine Specimen

While sorting through urine specimens in the toxicology section of the lab, a medical technologist noticed that one of the specimens had a faint pink color and a fruity odor. She took the sample to the toxicologist for instructions on what to do.

"What tests were ordered?" he asked.

"Methadone" she replied. "It's from the substance use clinic at Mercy Hospital."

"As I suspected. You can skip the immunoassay and run this on our LC-mass spec."

Analysis of the specimen gave the following results.

Analysis by LC-MS

Methadone	1500 ng/mL
EDDP	Not detected

- Why did the toxicologist insist on analyzing the specimen by LC-MS instead of the usual immunoassay screening test?
- Is this patient taking methadone?

Discussion

Methadone is prescribed for pain management and as opioid substitution therapy. Patients may divert their supply to friends and acquaintances or they may sell methadone to raise money for other drugs. Regular testing is done to ensure that patients are taking their medication.

Orally administered methadone is available in tablet or liquid form. Liquid preparations may be colored and flavored. An alert technologist noticed the unusual appearance and smell and the toxicologist, who had seen similar cases, suspected that methadone had been added directly to the urine sample.

People on methadone treatment excrete both the unchanged drug and its major metabolite, 2-ethylidene-1,5-dimethyl-3,3-diphenylpyrrolidine (EDDP). Screening immunoassays detect either methadone or EDDP, and results are reported as positive or negative. Analysis of samples with LC-MS gives quantitative results for the parent drug and its metabolite, so that adulterated samples can be identified.

The negative result for EDDP in this urine specimen indicates that the drug did not pass through a human body. The patient is not taking methadone, but he spiked a small amount into his urine to give the impression that he is compliant.

Takeaway messages

- People sometimes add drugs directly to urine specimens to show compliance with their treatment program.
- Measurement of the parent drug and its metabolite(s) can identify samples that are adulterated in this way.

Case 10.4 One Drug Too Many

SB, a 48-year-old woman, sustained fractures of two vertebrae in her cervical spine when her car was hit from behind by a delivery truck. She was treated with escalating does of oral morphine until the pain was under control. Six months later, she continues to take morphine tablets twice per day.

Her physician has been monitoring SB's medication use with occasional urine drug screens. The latest test came back positive for opiates, as expected. However, confirmation testing of urine gave the following results.

Analysis by LC-MS

Morphine	165 000 ng/mL
Hydromorphone	1700 ng/mL

At her next appointment, the physician showed the report to SB and asked if she was taking hydromorphone. The patient vigorously denied this, saying that she was following the doctor's orders and taking morphine exactly as prescribed.

- Is this patient taking hydromorphone as well as morphine?
- If not, how would you explain the positive result?

Discussion

Most morphine is metabolized by conjugation with glucuronic acid and then excreted in urine. When taken in large amounts or for long periods of time, a fraction of morphine is metabolized to hydromorphone. In this case, the amount of hydromorphone in urine is about 1% of the morphine concentration. The patient's claim that she is taking only morphine is therefore credible.

The presence of another opioid in the patient's urine was a warning sign of potential misuse. Some patients request prescriptions from multiple doctors or steal medications from family members or friends. Physicians may decide to discharge patients from their practice if they are seeking opioids from other sources.

Takeaway message
- Small amounts of hydromorphone may appear in urine specimens from patients on long-term morphine therapy.

Case 10.5   The Boss Wants to See You

An investment brokerage on Wall Street decided to set up an in-house drug testing program for its employees. This move followed a sensational story in the *New York Times* about drug abuse in the securities industry. The story had triggered investigations by the district attorney's office and led to indictments of several high-profile account managers at the firm.

The day after the announcement, all senior managers and floor traders were escorted to the executive suite on the 52nd floor. Security guards handed them a sample cup and led them individually to the lavatory. Urine samples were collected from each employee and shipped to Gold Dust Laboratories for analysis.

Seven employees tested positive for cocaine, two tested positive for amphetamines, and one tested positive for opiates. All received termination slips, but the opiate-positive employee challenged the test result. His urine sample was sent to a second laboratory for confirmation testing. The results were as follows.

Analysis by GC-MS

Codeine	6500 ng/mL
Morphine	700 ng/mL

The president of the company called the employee into his office, looked at the report and said, "Morphine … codeine … these are narcotics. Nobody working at my firm takes illegal drugs. You're fired!"

- What drug(s) has the employee been taking? Are they illegal?
- Is there a reasonable explanation for the positive drug test?

Discussion

The employee's test report indicates that he was taking codeine. Morphine is a metabolite of codeine – it is present at a lower concentration than the parent drug. Codeine is found in medications that are prescribed for pain relief and to control coughing. The pain reliever Tylenol® #3 contains 300 mg of acetaminophen, 15 mg of caffeine, and 30 mg of codeine in each tablet. Codeine is also a component of some cough syrups. Because codeine has less than one-fifth the potency of morphine, codeine-containing medications are available over the counter in certain jurisdictions. When purchased in this manner or by prescription, codeine is legal to possess and consume.

This employee was suffering from a cold and took a codeine-containing cough syrup the night before he was tested. It was prescribed to him by his family doctor – there was nothing illegal about it. His boss incorrectly assumed that he was abusing drugs.

Takeaway messages
- Morphine is a metabolite of codeine and is expected to appear in urine when someone takes codeine.
- Investigation of positive workplace drug tests is best done by a medical review officer to ensure that the process is handled in a fair and professional manner.

Epilogue
After leaving the firm, the employee went to law school, took a job in the district attorney's office, and eventually prosecuted the president who had fired him. The ex-president is now serving a 5-year sentence for securities fraud and tax evasion at a minimum-security prison in upstate New York.

Case 10.6   Dr Titus Saves the Day

The pathologist's phone rang at 1.30 in the afternoon. "Hello, Dr Titus speaking."

"Hi, this is Dr Sweeney. I'm a family practitioner, and I wanted to ask you about some test results that just came back from your lab. Two weeks ago, I saw a new patient. She has a herniated disc between L5 and S1 that causes her a lot of pain, and she has been taking oxycodone for several months now. When I refilled her prescription, she asked me to increase the dose from 30 to 60 mg/day."

"Several months of oxycodone treatment, and now she is asking you to double the dose," said Dr Titus. "Are you worried that she is becoming dependent on the drug?"

"Actually, that brings me to the point of this call," said Dr Sweeney. "When I tested her urine sample for opiates, the result was negative. I think she might be stockpiling or diverting the oxycodone."

"Was she ever tested for opioids before?" Dr Titus asked.

"I couldn't find anything in the records from her previous doctor. She seems like a reliable patient, but you never know . . ."

- How can you determine whether the patient is taking her medication?
- What advice would you give to this family doctor?

Discussion
Screening tests for opioids are usually performed with immunoassays. The opiates immunoassay can reliably detect morphine, codeine, and 6-acetylmorphine. It is less sensitive at identifying hydrocodone and hydromorphone, and oxycodone may not be detected at all (Table 10.2).

Dr Titus took the patient's report to the toxicology lab supervisor. She confirmed that the patient had been tested with the opiates immunoassay. Fortunately, the urine sample had not been discarded, and Dr Titus asked for an opioid profile by LC-MS. This showed the following.

Analysis by LC–MS

Oxycodone	3400 ng/mL
Oxymorphone	2800 ng/mL
Noroxycodone	680 ng/mL

Dr Titus called Dr Sweeney, gave her the results and said that the patient was indeed taking her oxycodone.

"Great," she responded. "That is so helpful!"

"Next time, I suggest that you order the oxycodone immunoassay screen," Dr Titus said. "It is quicker and easier than the LC-MS test we did today, and it's reliable. But I'm still concerned about the patient's request for a higher dose of oxycodone – it is an addictive drug. You may want to refer her to a pain management specialist."

Takeaway messages

- Oxycodone may not be detected by screening immunoassays for opiates. Instead, order oxycodone on the test requisition. The laboratory will choose an appropriate method of analysis.
- A negative result for oxycodone (or any prescribed medication) could indicate noncompliance by the patient. It raises suspicion that the drug is being diverted.
- Oxycodone and other prescription opioids are highly addictive. Requests for additional medication may be a warning sign that the patient is becoming dependent.

Further Reading

Articles

Armenian, P., Vo, K.T., Barr-Walker, J., and Lynch, K.L. (2018). Fentanyl, fentanyl analogs and novel synthetic opioids: a comprehensive review. *Neuropharmacology* 134: 121–132.

Milone, M.C. (2012). Laboratory testing for prescription opioids. *J Med Toxicol* 8: 408–416.

Saitman, A., Park, H.-D., and Fitzgerald, R.L. (2014). False-positive interferences of common urine drug screen immunoassays: a review. *J Anal Toxicol* 38: 387–396.

Book chapter

Nelson, L.S. and Olsen, D. (2015). Opioids. In: *Goldfrank's Toxicologic Emergencies*, 10e (eds. R.S. Hoffman, M.A. Howland, N.A. Lewin, et al.), 492–509. New York: McGraw-Hill.

Websites

National Institute on Drug Abuse
Fentanyl
   www.drugabuse.gov/publications/drugfacts/fentanyl
Heroin
   www.drugabuse.gov/publications/drugfacts/heroin
Prescription opioids
   www.drugabuse.gov/publications/drugfacts/prescription-opioids

Centre for Addiction and Mental Health
Buprenorphine
   www.camh.ca/en/health-info/mental-illness-and-addiction-index/buprenorphine
Fentanyl
   www.camh.ca/en/health-info/mental-illness-and-addiction-index/street-fentanyl
Heroin
   www.camh.ca/en/health-info/mental-illness-and-addiction-index/heroin
Methadone
   www.camh.ca/en/health-info/mental-illness-and-addiction-index/methadone
Naloxone
   www.camh.ca/en/health-info/mental-illness-and-addiction-index/naloxone
Prescription opioids
   www.camh.ca/en/health-info/mental-illness-and-addiction-index/prescription-opioids

Videos

Why the human brain loves opioids
   www.youtube.com/watch?v=fVdXlB89QOA

involved with thinking, attention, memory, emotion, and control of movement. CB2 receptors are found mainly on cells of the immune system (e.g., lymphocytes and monocytes).

Physiological Effects

The major effects of cannabinoids on the central nervous system are:

- euphoria
- relaxation
- feeling of well-being
- distorted perception of time
- enhanced or altered sensation
- loss of coordination
- impaired memory
- difficulty thinking and solving problems
- mood changes (e.g., anxiety)
- hallucinations, delusions (in high doses).

The effects listed above are for THC. Cannabidiol (CBD), the other major cannabinoid, does not have psychoactive properties. It binds to but does not activate the CB1 receptor, and it opposes some of the effects of THC.

Therapeutic Uses

Medicinal cannabis is legal in 36 states and the District of Columbia, and in Canada. THC reduces nausea and vomiting, stimulates the appetite, has analgesic properties, decreases spasticity, and lowers intraocular pressure. The most common conditions treated with medicinal cannabis are nausea and vomiting due to chemotherapy, weight loss in acquired immunodeficiency syndrome (AIDS), multiple sclerosis, and glaucoma.

Two oral cannabinoids, dronabinol (synthetic THC) and nabilone (a synthetic derivative of THC), are available in the United States and Canada.

Potential for Abuse

In spite of its approved therapeutic uses, which are governed by individual states, cannabis is listed as a Schedule 1 drug by the US Drug Enforcement Administration. Drugs in this class are defined as having a high potential for abuse and no currently accepted medical treatment use.

Cannabis is the most widely used illicit substance in the US. In 2018, more than 43 million people aged 12 years and older (15.9% of this population) reported use of cannabis in the past year. The rate was higher in males (18.5%) than females (13.4%), and cannabis use was highest among people aged 18–25 years (34.8%) and 26–34 years (29.6%).

Cannabis is consumed by smoking the dried plant in cigarettes (joints), pipes, or water-pipes (bongs). When smoked, the effects of cannabis are evident within a few minutes. Cannabis may also be mixed with food, such as brownies or cookies, or brewed as a tea. Orally consumed cannabis takes effect in about 30 minutes to an hour.

Chronic users may experience mild withdrawal symptoms, such as insomnia, anxiety, and irritability, when they discontinue the drug. The issue of cannabis as a gateway drug to other, more dangerous drugs is not settled. Most people who use cannabis do not go on to use other illicit substances.

Over the past decade, many synthetic cannabinoids have appeared on the market. These compounds are sprayed onto dried plant material so that they can be smoked, or they are available as liquids to be vaporized and inhaled. The products are sold under different brand names, such as Spice and K2. Synthetic cannabinoids produce some of the same pleasurable effects as cannabis, but they can cause harmful side-effects as well.

Metabolism

THC is metabolized mainly to 11-hydroxy-THC (11-OH-THC), which has biological activity, and then to the inactive 11-nor-9-carboxy-THC (THC-COOH; Figure 11.2). A graph showing the time course of THC metabolism and clearance from plasma appears in Figure 11.3. THC-COOH and its glucuronide conjugate are the main excretion forms in urine.

THC is a very lipophilic molecule and is stored in fat-containing tissues within the body. Slow diffusion out of cells can cause positive drug tests for days to weeks following the most recent use of cannabis.

Testing for Cannabinoids

Screening

Immunoassays for cannabinoids are designed to detect the major metabolite THC-COOH. For workplace drug testing, the cut-off for a positive result in urine is 50 ng/mL. A lower cut-off of 20 ng/mL may be used by some laboratories when testing clinical samples.

Figure 11.3 Mean plasma levels of THC, 11-OH-THC and THC-COOH during and after smoking a single 3.55% THC marijuana cigarette. *Source:* Reproduced from Huestis MA, Henningfield JE, Cone EJ. Blood cannabinoids. I. Absorption of THC and formation of 11-OH-THC and THC-COOH during and after smoking marijuana. J Anal Toxicol 1992;16:276–282, with permission from Oxford University Press.

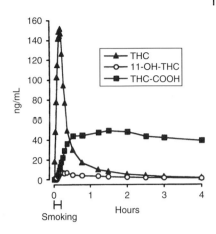

Table 11.1 Cross-reactivity of cannabinoids in a commercial immunoassay.

Cannabinoid	% Cross-reactivity
THC-COOH	100
8,11-$(OH)_2$-THC	86
8-OH-THC	74
11-OH-THC	65
THC-COOH-glucuronide	53

Source: Siemens ADVIA Chemistry Systems, Cannabinoid_2 Reagents (THC_2). Cross-reactivity was determined at the 50 ng/mL cut-off for a positive test.

Several other metabolites of THC are also detected in the screening immunoassay (Table 11.1). These compounds are present at lower concentrations than THC-COOH in urine specimens. However, it does underscore the fact that a mixture of cannabinoids may contribute to a positive screening result.

The issue of positive drug tests from inhaling second-hand smoke is often raised. Studies indicate that a cut-off of 50 ng/mL is sufficiently high to give a negative result in this group. Products made from hemp oil, including CBD preparations, may contain small amounts of THC. In some cases, consumers of these products may excrete enough THC-COOH to give a positive screening test for cannabinoids. If there is any question about a positive result, the patient should be advised to discontinue the product and be retested at a later time.

Synthetic cannabinoids do not give a positive screening test with immunoassays, as their structures are different from the naturally occurring cannabinoids.

Confirmation

The threshold for confirming a positive screening test is 15 ng/mL of THC-COOH. Either gas chromatography-mass spectrometry (GC-MS) or liquid chromatography-tandem mass spectrometry (LC-MS/MS) is typically used for confirmation testing. The psychoactive compound THC can also be detected by these techniques. Identifying THC is useful in legal cases (e.g., driving under the influence of cannabis), but it is not necessary in most clinical situations.

One or more of the synthetic cannabinoids may be detected by MS-based techniques if an assay specific for that compound has been developed.

Window of Detection

Cannabinoids can be detected in urine for up to 3 days following a single use, up to 10–15 days with regular (daily) use, and up to a month or more with long-term heavy use. The length of time depends upon the THC content of smoked cannabis, frequency and duration of use, hydration state of the patient, and analytical cut-off of the assay.

Interferences

Efavirenz (an antiviral drug for treating HIV infection), nonsteroidal antiinflammatory drugs (NSAIDs), and proton pump inhibitors have been reported to cross-react in immunoassays for cannabis and give false-positive results. Patients taking dronabinol, a synthetic form of THC, may test positive as well.

Case Studies

Case 11.1   Busted

While on patrol one evening, Officer Friday noticed a car with an expired license plate weaving across the center line of a downtown street. He activated the flashing lights on top of his cruiser, signaled the driver to pull over, and approached the car on foot. Inside were two young men wearing identical green baseball caps with the words "Cannabis World" embroidered on the front. A heavy metal song was blaring on the radio, and the smell of cannabis smoke was pungent.

Officer Friday asked the driver to step out of the car. He conducted a standardized field sobriety test to assess balance, coordination, and eye movements. The driver followed instructions and thought he had successfully completed all three tasks. However, Officer Friday concluded that the driver was intoxicated and took him to the police station, where blood was drawn for several tests. Results were as follows.

Analysis by GC

Ethanol	<10 mg/dL (<2 mmol/L)

Analysis by LC-MS

THC	9 ng/mL
THC-COOH	110 ng/mL

- Is driving under the influence of cannabis against the law?
- Would you consider this driver to be legally intoxicated?

Discussion
Driving under the influence of cannabis is a criminal offence. Cannabis slows reaction time, impairs coordination, and interferes with concentration, all of which are important when operating a motor vehicle. Drivers who have smoked cannabis can't respond quickly to changing road and traffic conditions. They tend to drive at varying speeds.

The definition of driving under the influence depends upon where a person lives. Laws are generally of three types.

- Zero tolerance law – prohibits driving with any amount of THC and/or its metabolites in the body.
- Per se law – prohibits driving with a blood concentration of THC that exceeds the legal limit.
- Under the influence – requires the driver to be under the influence of or affected by THC.

Among states with per se limits, a blood THC concentration of 1, 2, or 5 ng/mL is considered the legal threshold. Current information can be found on the National Conference of State Legislatures website: www.ncsl.org/research/transportation/drugged-driving-overview.aspx

In Canada, driving with at least 2 ng/mL of THC in blood is prohibited, and the penalties escalate for blood THC levels of 5 ng/mL or more. Details are provided on the Department of Justice website of the Government of Canada: www.justice.gc.ca/eng/cj-jp/sidl-rlcfa/

The driver in this case had a blood THC concentration above the legal limit and, according to Officer Friday, was showing signs of intoxication. He would meet the legal criteria for driving under the influence of cannabis in any state or province.

Takeaway messages
- Smoking or otherwise consuming cannabis impairs driving ability.
- The degree of impairment at a given THC blood level varies widely among individuals.
- Legal intoxication is determined in part or in whole by measurement of THC and its metabolites in blood. Laws are not consistent across different states.

Case 11.2 The Cannabidiol Conundrum

SH is a 57-year-old woman with long-standing arthritis in her hands and wrists. Recently a naturopath recommended that she take an oral CBD preparation every day and apply a topical CBD-containing cream to affected joints as needed. After trying this therapy for 3 weeks, the pain in her hands practically disappeared, and her range of motion improved.

At her next visit to the naturopath, SH was thrilled with the results, but she did have one concern.

"I work for an aerospace company that has contracts with the Defense Department – every employee is tested for drugs once or twice a year. Will CBD cause the drug test for cannabis to turn positive?"

- How would you answer her question?
- Would you change her therapy or suggest other ways to deal with this issue?

Discussion
CBD comes from *Cannabis sativa*, the same plant that produces THC. There are two varieties of the plant: hemp, which contains <0.3% THC, and marijuana, which has 5–30% THC. CBD can be extracted from either plant – the amount of THC will be higher in extracts from the marijuana plant.

CBD and its metabolites are not detected in the screening immunoassay for cannabis. The test is designed to detect THC-COOH, the main metabolite of THC. Since some CBD products contain more THC than others, it is possible that using these products will cause a positive test for cannabis.

People use CBD to treat pain, anxiety, depression, insomnia, and other ailments. The number of scientific studies on CBD is too limited to draw conclusions on the effectiveness of this treatment. However, many patients find relief from using CBD.

If SH wants to continue taking CBD for her arthritis pain, she has several options. First, she can look for CBD products made from hemp to ensure that the THC content is low. Second, she can advise her employer that she takes CBD daily for arthritis, and that it may cause a positive cannabis drug test.

Takeaway messages
- People who use CBD-containing products may test positive for cannabis on urine drug screens.
- It is better to advise an employer of CBD use beforehand, rather than trying to explain a positive test after the fact.

Case 11.3 An Olympic Gold Medal

In 1998, Canadian snowboarder Ross Rebagliati raced down the slopes of a giant slalom course and into the history books. Snowboarding debuted as an Olympic sport at the Winter Olympics that year in Nagano, Japan, and Ross won the first gold medal in a snowboarding event. After the competition, a urine drug test revealed evidence of cannabis use, and he was stripped of the gold medal by the International Olympic Committee. Less than two days later the medal was reinstated, because cannabis was not on the list of banned substances.

Rebagliati said that he did not smoke cannabis during the competition. He claimed that the traces of cannabis in his system were due to a significant amount of time spent in an environment with marijuana users. According to published reports, his urine contained 17.8 ng/mL of the THC metabolite.

- Is cannabis a performance-enhancing drug?
- Can athletes compete if there is evidence of cannabis use?
- Does secondhand cannabis smoke ever cause a positive urine drug test?

Discussion
There are differing views on whether cannabis should be considered a performance-enhancing drug. Smoking cannabis before a competition may relieve pain, reduce anxiety, and allow the athlete to focus. On the other hand, cannabis produces limb fatigue, impairs motor control and balance, and slows reaction time.

The World Anti-Doping Agency (WADA) placed cannabis on its list of prohibited substances in 2004. To be placed on this list, a substance must meet at least two of the following three criteria:

- has the potential to enhance performance
- poses a health risk to athletes
- violates the spirit of sport.

Athletes are not allowed to use cannabis while in competition, but there are no restrictions on recreational cannabis use outside competition. The threshold value in urine for a positive test is 150 ng/mL of THC-COOH. This cut-off was selected to differentiate recent cannabis use from intake that occurred days or weeks earlier. CBD is the only cannabinoid, natural or synthetic, that is not on the list of prohibited substances.

Experiments with passive exposure to cannabis smoke have shown that urine THC-COOH values rarely reach 15 ng/mL. The screening immunoassay for cannabis (with a 50 ng/mL cut-off) is virtually never positive from secondhand smoke except under extreme smoke conditions in an unventilated room.

Takeaway messages
- Cannabis use outside competition does not disqualify athletes from taking part in sporting events.
- Secondhand smoke is very unlikely to cause a positive test for cannabis in urine, either at the screening (50 ng/mL) or confirmation (15 ng/mL) level.

Case 11.4   Dr Titus Saves the Day – Again

Dr Titus, the medical director of the toxicology laboratory, was contacted by Sandy, a nurse working at an addiction treatment facility. She wanted to discuss results she had just received on one of her patients.

"I hope you can help me with these test results. I have a fairly new patient with cannabis use disorder who says he wants to get clean. We collect a urine sample every week to test for ongoing cannabis use. This is the third test in a row to come back positive!"

"Yes" Dr Titus said, "I'm looking at the results right now and can see the positive drug screens."

"The patient seems so serious about quitting, and he swears that he has not touched a joint since entering the program. Do you think that he is still smoking cannabis? Our meetings become really tense when I bring up his test results, and he needs support. I'm worried that he will just give up."

"How long has he used cannabis?" asked Dr Titus.

"Years" said Sandy. "He smoked several joints a day ... at least."

"And when did he quit?"

"He says he took his last puff 4 weeks ago, just before coming to our treatment center."

- Do the test results indicate recent use of cannabis, or is there another explanation?
- Is there anything more that the lab can do?
- What advice should Dr Titus give to Sandy?

Discussion

People who use cannabis over long periods of time can continue to show positive urine tests for a month or longer. THC is a very lipophilic molecule – it distributes into body fat and is slowly released, metabolized, and excreted. The three positive test results could be due to washout of the accumulated THC in this patient. A negative screening test for cannabis prior to these positive tests would indicate ongoing consumption – but we do not have that evidence.

The immunoassay screening test for cannabis is very reliable, so confirmation testing is not usually necessary. However, confirmation tests provide a quantitative value of THC-COOH in urine. If the patient had indeed stopped consuming cannabis, those values would be expected to decrease with each passing week.

Dr Titus located the two most recent urine specimens and asked the senior technologist to analyze them by LC-MS. The results for THC-COOH were as follows.

Analysis by LC-MS

October 12	95 ng/mL
October 19	60 ng/mL

He called Sandy back and reported these values. "In my opinion," he told her, "the patient is still excreting THC that was consumed before he started treatment. When he says it's been 4 weeks since his last puff, I believe him."

Takeaway messages

- Cannabinoids are excreted slowly. Long-term users may continue to test positive for a month or longer after last using cannabis.
- Quantitative tests for THC-COOH at two or more time points can help to distinguish recent use from more remote use.

Further Reading

Articles

Kulig, K. (2017). Interpretation of workplace tests for cannabinoids. *J. Med. Toxicol.* 13: 106–110.

Mills, B., Yepes, A., and Nugent, K. (2015). Synthetic cannabinoids. *Am. J. Med. Sci.* 350: 59–62.

Whiting, P.F., Wolff, R.F., Deshpande, S., et al. (2015) Cannabinoids for medical use: A systematic review and meta-analysis. *JAMA* 313: 2456–2473.

Book chapter

Lapoint, J.M. (2015). Cannabinoids. In: *Goldfrank's Toxicologic Emergencies*, 10e (eds. R.S. Hoffman, M.A. Howland, N.A. Lewin, et al.), 1042–1053. New York: McGraw-Hill.

Websites

National Institute on Drug Abuse
Marijuana
   www.drugabuse.gov/publications/drugfacts/marijuana
Synthetic cannabinoids (K2/Spice)
   www.drugabuse.gov/publications/drugfacts/synthetic-cannabinoids-k2spice

Centre for Addiction and Mental Health
Cannabis
   www.camh.ca/en/health-info/mental-illness-and-addiction-index/cannabis

State Medical Marijuana Laws, National Conference of State Legislatures
   www.ncsl.org/research/health/state-medical-marijuana-laws.aspx

Videos

Visualization of the endocannabinoid signaling system
   www.youtube.com/watch?v=jznQfMj9RWM
Cannabis and cannabinoids
   www.youtube.com/watch?v=lkNIRZXraY4

AACC Pearls of Laboratory Medicine
Synthetic drugs: cathinones and cannabinoids
   www.aacc.org/science-and-research/clinical-chemistry-trainee-council/trainee-council-in-english/pearls-of-laboratory-medicine/2015/synthetic-drugs-cathinones-and-cannabinoids

12

Hallucinogens

Hallucinogens are drugs that alter an individual's perception of reality. They can produce distortions of what is sensed (illusions), and they can create false perceptions of the external environment (hallucinations). Hallucinogens also affect mood, thought, and judgment. Some are derived from natural sources, while others are synthetic.

Hallucinogens are often divided into two categories: classic hallucinogens and dissociative drugs.

Classic Hallucinogens

The classic hallucinogens include lysergic acid diethylamide (LSD), psilocybin, and mescaline.

Lysergic Acid Diethylamide (LSD)

LSD is a derivative of lysergic acid, an alkaloid that occurs naturally in the fungus *Claviceps purpurea*. The fungus, which grows on rye and other grains, is also known by the common name ergot, and it produces a number of pharmacologically active compounds called ergot alkaloids. LSD was first synthesized in 1938 by the chemist Albert Hofmann. Its hallucinogenic properties were discovered 5 years later when he intentionally took a dose of the drug. The d-isomer is a potent hallucinogen.

LSD became popular in the 1960s as a drug associated with the counterculture movement. The music, art, and literature of the time contained both implicit and explicit references to the psychedelic effects of the drug.

A portion of the LSD molecule resembles serotonin, one of the major neurotransmitters within the central nervous system (Figure 12.1).

An Introduction to Testing for Drugs of Abuse, First Edition. William E. Schreiber.
© 2022 John Wiley & Sons Ltd. Published 2022 by John Wiley & Sons Ltd.

Psilocybin

Psilocybin is produced by mushrooms of the genus *Psilocybe* (Figure 12.2). Following ingestion, the phosphate group is removed to release psilocin, the active form of the drug. The structure of psilocin is similar to that of serotonin, differing only in the position of the hydroxy group on the aromatic ring and the presence of two methyl groups on the amine (Figure 12.1).

   Psilocybin-containing mushrooms have been used for centuries by native peoples, often as part of religious ceremonies. Depending upon the species of mushroom, the amount of psilocybin varies between 0.1% and 1.5% of the dried weight. A typical dose of psilocybin is 10–50 mg, which corresponds to about 20–30 g of fresh mushrooms.

Mescaline

Mescaline is an alkaloid found in the peyote cactus, *Lophophora williamsii*, as well as other members of the cactus family (Figure 12.3). Its structure, which is based on phenylethylamine, is similar to methylenedioxymethamphetamine (MDMA) and other amphetamines.

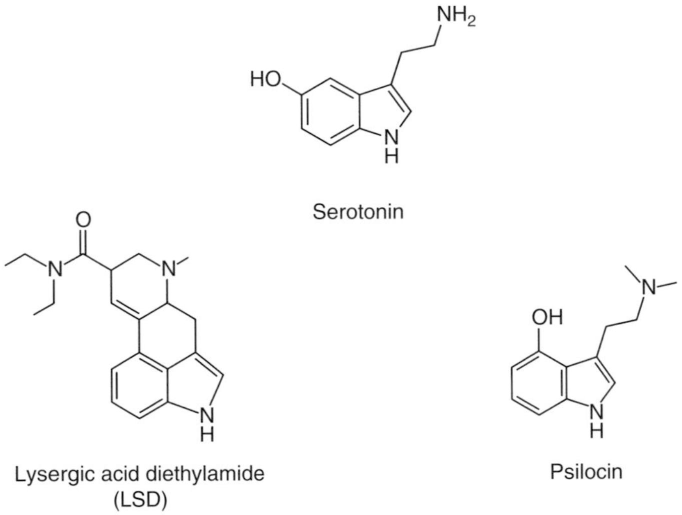

Serotonin

Lysergic acid diethylamide
(LSD)

Psilocin

Figure 12.1 Structures of the neurotransmitter serotonin and the hallucinogenic agents lysergic acid diethylamide (LSD) and psilocin. Both drugs bind to a specific type of serotonin receptor (5-HT$_{2A}$).

Figure 12.2 *Psilocybe cubensis,* a species of psilocybin-containing mushrooms. *Source:* Yarygin/Shutterstock.

(a)

(b)

Figure 12.3 (a) Structure of mescaline (3,4,5-trimethoxyphenylethylamine). (b) *Lophophora williamsii,* the peyote cactus. *Source:* Vainillaychile/Shutterstock.

Peyote has been consumed by natives of the southwestern United States and northern Mexico since the beginning of recorded history. The top of the cactus is cut from the roots, dried, and either chewed or soaked in water to make a tea. A single dose of mescaline is 300–500 mg, equivalent to about 5 g of dried peyote cactus.

How Hallucinogens Work

LSD, psilocybin, and mescaline exert their perception-altering effects by acting on neural circuits in the brain that use serotonin as the neurotransmitter.

Serotonin mediates many functions – mood, learning, memory, perception, appetite, sexual behavior, regulation of temperature, sleep, and more. After serotonin is released from nerve endings, it binds to and activates serotonin receptors on adjacent neurons.

The serotonin receptor family is divided into seven major types and even more subtypes. One of them, the 5-HT$_{2A}$ receptor, has the greatest affinity for the hallucinogens. This receptor is expressed in high concentration on neurons of the cerebral cortex. When a hallucinogenic drug binds to the 5-HT$_{2A}$ receptor, it excites the corresponding neuron and stimulates transmission of a nerve impulse.

Some psychiatric disorders are linked to changes in serotonin, and drugs that affect serotonin activity (e.g., antidepressants, antipsychotics) are used in their treatment.

Physiological Effects

People who take hallucinogens may see images, hear sounds, and feel sensations that do not exist, although they seem very real. Other short-term effects of these drugs include:

- alterations in sensory perception
- distorted sense of time
- relaxation, panic, or paranoia
- psychotic thinking, bizarre behavior
- increased heart rate and blood pressure
- hyperthermia, sweating
- nausea
- dry mouth
- loss of appetite
- lack of coordination
- changes in sleep.

A recurrence of hallucinations or other sensory experiences (flashbacks) may occur days, months or even years after taking the drug. In rare cases, the drug user may develop a persistent psychosis with visual disturbances, changes in mood, paranoia, and/or disordered thought patterns.

Therapeutic Uses

All three of the classic hallucinogens are listed as Schedule I drugs by the US Drug Enforcement Administration. There are no accepted medical uses for any of these substances.

Potential for Abuse

Hallucinogens differ from other drugs of abuse in that they do not induce dependence or addiction, and there is no withdrawal syndrome. However, tolerance develops with repeated use.

In 2018, an estimated 5.6 million people in the United States aged 12 years and older (2.0% of this population) reported use of hallucinogens in the past year. The highest rate of use was among people aged 18–25 years (6.9% of this cohort). LSD, mescaline, psilocybin, phencyclidine (PCP), ketamine, and several other drugs or plant preparations were considered hallucinogens for the purposes of the survey.

The classic hallucinogens produce physical signs and symptoms, but they rarely if ever cause death. However, people who consume these drugs may act on their altered perceptions and place themselves in danger, sometimes with fatal consequences.

Metabolism

The major metabolite of LSD is 2-oxo-3-hydroxy-LSD. Small amounts (<5%) of the unchanged drug are also excreted in urine. More than half of a dose of mescaline is excreted unchanged, and most of the remainder is converted to a carboxylic acid derivative. Psilocybin is rapidly converted to the biologically active psilocin in vivo. Psilocin is further metabolized to 4-hydroxyindoleacetic acid or conjugated to glucuronic acid prior to excretion. Elimination half-lives for these drugs appear in Table 12.1.

Testing for Hallucinogens

A screening immunoassay for LSD is available. The cut-off for a positive result is 0.5 ng/mL. Lower concentrations (down to 0.1 ng/mL) can be detected by liquid chromatography-mass spectrometry (LC-MS). Assays for mescaline and psilocin are available at specialized toxicology laboratories and are based on gas chromatography-mass spectrometry (GC-MS) or LC-MS techniques.

Table 12.1 Elimination half-life data for hallucinogens.

Drug	Half-life (h)
LSD	3–4
Mescaline	6
Psilocin	1.8–4.5
Phencyclidine	7–46
Ketamine	3–4

Source: Baselt RC. Disposition of Toxic Drugs and Chemicals in Man, 9th edn. Seal Beach, CA: Biomedical Publications, 2011.

Window of Detection

All the hallucinogens have short half-lives, and this limits the time frame for detection. Screening assays can detect LSD in urine up to a day after use. Mescaline is detectable for 2–3 days following ingestion, and psilocin may be present for 1–3 days, depending upon the sensitivity of the assay.

Interferences

A number of drugs, many of them antidepressants or antipsychotics, can give false-positive screening results in the immunoassay for LSD. A positive screening test should be confirmed by GC-MS or LC-MS analysis.

Dissociative Drugs

This group of drugs can alter sensory perceptions and cause hallucinations, but in addition they create a sense of detachment or dissociation from reality.

Phencyclidine (PCP)

PCP is a synthetic compound that was commercialized as an anesthetic agent in the 1950s. It was removed from the market in 1965 due to its adverse side-effects and the availability of better anesthetic drugs, such as ketamine. The structure consists of three rings – phenyl, cyclohexane, and piperidine – bound into a single molecule (Figure 12.4).

Figure 12.4 Structures of the dissociative drugs phencyclidine and ketamine.

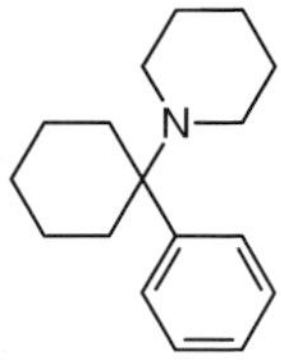

Phencyclidine (PCP) 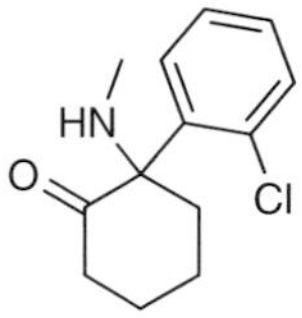   Ketamine

Ketamine

Ketamine was first synthesized in 1962 and approved for use as an anesthetic in 1970. The drug has structural similarities to PCP (Figure 12.4) but is missing the piperidine ring.

How Dissociative Drugs Work

PCP and ketamine produce their effects by interfering with the activity of the neurotransmitter glutamate. Glutamate-dependent neural pathways are involved in learning and memory, emotion, and perception of pain.

Following its release from nerve endings, glutamate binds to the N-methyl-D-aspartate (NMDA) receptor, which is located on the postsynaptic membrane of adjacent neurons. Binding of glutamate opens an ion channel that allows cations to flow across the membrane. This causes depolarization of the neuron and propagation of the nerve impulse.

PCP and ketamine both bind within the ion channel pore and restrict the flow of cations into the cell. The resulting inhibition of nerve signaling produces analgesia and anesthesia and may contribute to the psychoactive properties of these drugs.

Physiological Effects

The mind-altering effects of PCP and ketamine are hallucinations, a sense of detachment from the environment, delusions, anxiety, and paranoia. Other signs and symptoms include:

- slurred speech
- loss of coordination
- increased heart rate, respiratory rate, and blood pressure
- nausea and vomiting
- dizziness
- blurred vision

- seizures
- coma.

Long-term use can cause memory loss, difficulty thinking, and loss of appetite.

Therapeutic Uses

PCP is listed as a Schedule II drug in the United States and is not available by prescription. Ketamine is used in general anesthesia, either by itself or in combination with other drugs. It is also administered to manage acute and chronic pain syndromes. Esketamine, the S-enantiomer of the drug, was recently approved for treatment of depression.

Potential for Abuse

PCP was a popular drug of abuse during the 1970s but is much less frequently encountered today. It is consumed for its hallucinogenic properties and the feelings of strength, invulnerability, and euphoria that it can produce. Overdoses can be lethal, but death among PCP users is usually due to dangerous behavior caused by the drug's mind-altering effects.

Ketamine induces a dream-like state and a sensation of floating outside the body. It is one of several club drugs taken by young people at nightclubs, dance parties, and concerts to stimulate the senses and enhance social interaction. These same drugs have been used in cases of drug-facilitated sexual assault ("date rape"). A limited number of ketamine-related fatalities have been reported, most of which involved other drugs or alcohol as well.

Metabolism

PCP is hydroxylated to form inactive metabolites, which are then conjugated to glucuronic acid and excreted in urine. About 10% of the drug is excreted unchanged. Ketamine is demethylated to norketamine, which also has biological activity. Both compounds appear in urine in small amounts (<5% of a dose). The half-lives of PCP and ketamine are given in Table 12.1.

Testing for Dissociative Drugs

Screening for PCP is performed by immunoassay. A concentration of 25 ng/mL or above is considered positive. Ketamine is not usually included in routine drug screens. Confirmation testing for both compounds as well as the metabolite norketamine is by GC-MS or LC-MS analysis.

Table 12.2 Drugs that can give a false-positive immunoassay result for phencyclidine.

Compound
Dextromethorphan
Diphenhydramine
Doxylamine
Ibuprofen
Imipramine
Ketamine
Lamotrigine
Methylenedioxypyrovalerone
Meperidine
Mesoridazine
Thioridazine
Tramadol
Venlafaxine, O-desmethylvenlafaxine

Source: Moeller KE, Kissack JC, Atayee RS, et al. Clinical interpretation of urine drug tests: what clinicians need to know about urine drug screens. Mayo Clin Proc 2017;92:774–796.

Window of Detection

PCP can be detected up to 8 days following casual use and longer in chronic users. The detection window for ketamine is about 2 days.

Interferences

Drugs that can give a false-positive screening result for PCP are listed in Table 12.2.

Case Studies

Case 12.1   Where's Dixon?
The head of the anesthesiology service at a large teaching hospital was preparing for the first case of the day, an elective cholecystectomy. The third year resident assigned to that operating room had not arrived yet. *(Continued)*

"Where's Dixon?," the anesthesiologist asked one of the nurses. "Can you find him? I think he was on call last night."

The nurse walked through the hallways adjacent to the operating rooms, looked in the call room to see if he had overslept, had him paged on the overhead loudspeakers and dialed his cell phone, all without success. She opened the door to the medication room and saw Dr Dixon lying on the floor behind a crash cart. He appeared dazed and unaware of his surroundings. She did a quick check of his pulse and respirations, then called down the corridor for help.

The anesthesiology resident was hoisted onto a wheeled stretcher and transported to the emergency department. Twenty minutes later he was able to converse with the doctor, a senior resident in emergency medicine. Dr Dixon claimed to be in excellent health, denied any medical problems, and said that he likely passed out after a difficult night on call with very little sleep.

"I hear you," said his colleague, "but it took a while for you to come around. Let's get some tests done just to be sure there's nothing else going on."

Two tubes of blood were drawn and a urine sample was collected. All of the results on the chemistry and hematology panels were normal. However, the emergency medicine resident was not satisfied with Dixon's explanation of events. Earlier that month, he had attended a grand rounds presentation about drug use among physicians. He decided to submit the samples to the toxicology laboratory for a comprehensive drug screen.

The next day, he received a phone call from the lab.

"We analyzed the samples that you sent down. They were negative for all the usual drugs of abuse, but we found traces of ketamine and norketamine in serum, and they were both clearly present in urine."

- How common is substance use among physicians?
- Which specialties have a higher prevalence of substance use?
- What substances are most frequently involved?
- Why did the anesthesiology resident take ketamine?

Discussion
About 10–15% of physicians will have a substance use disorder at some point during their lifetime. This figure is similar to the prevalence of substance use disorders in the general population. The specialties with the highest rates of substance use are anesthesiology, emergency medicine, and psychiatry – pediatrics has the lowest rate. Alcohol is the most frequently used substance, followed by prescription opioids.

The reasons for substance use by physicians are often the same as those given by nonmedical people. Physicians have additional risk factors that are unique to their profession. These include greater access to prescription drugs, high expectations from their patients and their peers, and a culture of self-reliance. Many physicians with substance use problems do not feel comfortable asking for help and attempt to hide their addiction.

A study of residents who entered anesthesiology programs from 1975 to 2009 showed that slightly less than 1% had evidence of a substance use disorder during training (Warner DO et al. JAMA 2013;310:2289–2296). The most commonly used substance at the initial episode was intravenous opioids. Other substances used (in descending order of frequency) were alcohol, cannabis or cocaine, anesthetics/hypnotics, and oral opioids. In the anesthetics category, propofol was more frequently used than ketamine or inhaled anesthetics.

We don't know why Dr Dixon chose to self-administer ketamine. Access to the drug in his work environment was likely a contributing factor.

Takeaway messages

- Substance use disorders affect physicians as well as members of the general population. The prevalence in both groups is similar.
- The high expectations placed on doctors make it difficult for them to admit that they have a substance use problem and seek help.
- Physicians have access to drugs that do not show up on routine screening tests.

Case 12.2   We Don't Offer that Test

Saturday morning, 2 a.m. Dr Titus was woken up by the phone ringing on his nightstand.

"Hello"?

"This is Dr Nighthawk in Emergency," a loud voice said. "Are you the pathologist on call? I've got three teenagers here who may have taken LSD."

"What"?

"They're spaced out, staring at the walls and talking to people who aren't there. They were at a concert downtown when they freaked out. A friend dropped them off at the ER."

"Who is this"?

"Nighthawk! Look, these kids are acting really strange, and I need to find out why. Can you run an LSD test for me"?

(Continued)

"It's two in the morning and there's only a skeleton crew on duty in the lab. Did you ask them if they took any drugs"?

"Yes – they swallowed some green pills that were being passed around at the concert."

"Okay," said Dr Titus. "I'll call the lab and ask for a targeted drug screen of urine. It's the best I can do until tomorrow."

"What about LSD?"

"We don't offer a test for LSD. Goodnight, Dr Hawk."

- Which drugs can cause hallucinations?
- Why doesn't the laboratory test for LSD?

Discussion

In addition to the classic hallucinogens (LSD, psilocybin, and mescaline) and dissociative drugs (PCP and ketamine), a number of other substances can cause hallucinations.

- Methylenedioxymethamphetamine (MDMA)
- Cannabis
- Dextromethorphan
- *Salvia divinorum*
- N,N-dimethyltryptamine (DMT)
- 25I-NBOMe

MDMA and cannabis are discussed in other chapters. Both substances can produce hallucinations when taken in large doses.

Dextromethorphan is a cough suppressant in some over-the-counter cough and cold medications. *Salvia divinorum* is a plant that grows in southern Mexico, Central and South America. It is consumed by chewing or smoking the leaves. DMT occurs naturally in a number of plants from the Amazon basin. These plants are brewed into a hallucinogenic tea called ayahuasca. 25I-NBOMe is a synthetic hallucinogen that was developed for neuroscience research but is also used recreationally.

Intoxication with hallucinogens is usually self-limiting with no residual effects after the drug wears off. Having a rapid test to identify these drugs would be convenient, but it is not necessary to properly manage patients. Screening immunoassays for PCP, MDMA, and cannabis are on the laboratory's test menu. Assays for the other drugs, including LSD, are available only in specialized toxicology laboratories.

The results of the screening tests were the same in all three patients.

Immunoassay

Amphetamines	Negative
Cannabis	Negative
Phencyclidine	Negative

The following week, Dr Titus sent the samples to a reference toxicology laboratory for further analysis. The results came back positive for LSD – Dr Nighthawk's suspicions proved correct.

Takeaway messages
- With the exception of PCP, testing for hallucinogens is not available in most clinical laboratories. Patients with an acute intoxication can be successfully managed without this information.
- Other substances that are not usually considered to be hallucinogens – MDMA, cannabis, and dextromethorphan – can cause hallucinations when taken in large amounts.

Further Reading

Article

Morgan, C.J.A. and Curran, H.V. (2011). Ketamine use: a review. *Addiction* 107: 27–38.

Book chapters

Carey, J.L. and Babu, K.M. (2015). Hallucinogens. In: *Goldfrank's Toxicologic Emergencies*, 10e (eds. R.S. Hoffman, M.A. Howland, N.A. Lewin, et al.), 1114–1123. New York: McGraw-Hill.

Olmedo, R.E. (2015). Phencyclidine and ketamine. In: *Goldfrank's Toxicologic Emergencies*, 10e (eds. R.S. Hoffman, M.A. Howland, N.A. Lewin, et al.), 1144–1154. New York: McGraw-Hill.

Websites

National Institute on Drug Abuse
Hallucinogens
   www.drugabuse.gov/publications/drugfacts/hallucinogens
Hallucinogens and dissociative drugs
   www.drugabuse.gov/publications/research-reports/hallucinogens-dissociative-drugs

Centre for Addiction and Mental Health

Hallucinogens

www.camh.ca/en/health-info/mental-illness-and-addiction-index/hallucinogens

Ketamine

www.camh.ca/en/health-info/mental-illness-and-addiction-index/ketamine

LSD

www.camh.ca/en/health-info/mental-illness-and-addiction-index/lsd

Videos

2-Minute Neuroscience: LSD

www.youtube.com/watch?v=WKPMOXD4s-c

2-Minute Neuroscience: Psilocybin

www.youtube.com/watch?v=XBEas8MGzd0

Your Brain on LSD and Acid

www.youtube.com/watch?v=wG5JyorwYPo

The first modern images of a human brain on LSD

www.youtube.com/watch?v=kvalFfavNpU

13

Alcohols

Alcohols are found in products that we use every day – disinfectants, cleaning solutions, automotive fluids, solvents and, perhaps most conspicuously, beverages. The four alcohols discussed in this chapter are shown in Figure 13.1.

Ethanol is considered to be a substance of abuse, due to its widespread consumption and its effects on human health and behavior. Methanol, ethylene glycol, and isopropanol are not safe to drink, as they can cause serious injury and even death. Nevertheless, thousands of people are poisoned by these compounds every year.

Ethanol

According to historical records, alcoholic beverages have been produced and consumed by humans since 6000 BC, and possibly earlier. The alcohol in these beverages is ethanol, a 2-carbon alcohol that is generated during the fermentation of sugars by yeast (Figure 13.1). The common types of alcoholic beverages and their typical alcohol content by volume are beer (4–7%), wine (6–16%), and distilled spirits (35–50%).

Many people find the effects of alcohol to be pleasurable. Social pressure and group norms also affect drinking behavior. When consumed in moderate amounts, the effects of alcohol on physical and mental health can be favorable. Heavy consumption reveals the toxic nature of alcohol on both individuals and society.

The term "alcohol" is often used as a synonym for ethanol – both terms will be used interchangeably in the first part of this chapter.

An Introduction to Testing for Drugs of Abuse, First Edition. William E. Schreiber.

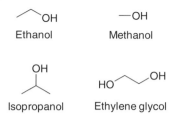

Ethanol

Methanol

Isopropanol

Ethylene glycol

Figure 13.1 Structures of ethanol, methanol, isopropanol, and ethylene glycol. The latter three compounds are sometimes consumed by people in place of ethanol. Methanol and ethylene glycol are converted to toxic intermediates that cause metabolic abnormalities, tissue injury and, in severe cases, death.

How Alcohol Works

The amount of alcohol required to produce a pharmacological effect is much greater than for other drugs of abuse. Blood alcohol concentrations are measured in mmol/L, whereas the concentrations for other psychoactive drugs are in the μmol/L or even nmol/L range, a difference of at least 1000-fold.

Ethanol has complex effects on physiology and behavior. Although it acts as a central nervous system (CNS) depressant, it also produces a mild degree of stimulation. The mechanisms by which alcohol exerts these effects are not fully understood. CNS pathways that involve the neurotransmitters gamma-aminobutyric acid (GABA), glutamate, and dopamine appear to be important in mediating those effects.

GABA is the main inhibitory neurotransmitter in the brain. When it binds to the GABA-A receptor, chloride ions flow into the neuron, which causes hyperpolarization and prevents the neuron from firing. Ethanol has been shown to modulate the activity of the GABA-A receptor and increase chloride flow into the cell, as do benzodiazepines and barbiturates. This process enhances the inhibitory effects of GABA.

Glutamate is the major excitatory neurotransmitter in the CNS. Following release into a synapse, glutamate binds to and activates receptors on postsynaptic neurons. Ionotropic receptors for glutamate, such as the N-methyl-D-aspartate (NMDA) receptor, contain ion pores that allow cations to enter the cell and stimulate an action potential. Ethanol inhibits the activity of the NMDA receptor, reduces the flow of cations into the cell, and decreases conduction of the nerve signal.

Dopamine functions as a neurotransmitter in the mesolimbic pathway of the brain. This circuit connects the ventral tegmental area to the nucleus accumbens, a portion of the brain that is involved in motivation and reward. Ethanol stimulates the pathway and drives dopamine release, which is thought to mediate the pleasurable effects of alcohol.

Physiological Effects

The signs and symptoms of acute ethanol toxicity are directly related to its concentration in blood (Table 13.1). Effects are more pronounced in people who consume alcohol infrequently or not at all than in those who drink regularly. Drivers with a blood alcohol concentration of 80 mg/dL (17 mmol/L) or higher

Table 13.1   Signs and symptoms of alcohol intoxication.

Blood alcohol concentration	Clinical features
30–120 mg/dL (6–26 mmol/L)	Mild euphoria Increased self-confidence Lowered inhibitions Diminished attention, judgment, self-control Limited sensorimotor impairment
90–250 mg/dL (20–54 mmol/L)	Loss of judgment Emotional instability Impaired perception, memory Decreased sensory response Slow reaction time Impaired balance, coordination Decreased visual acuity Drowsiness
180–300 mg/dL (39–65 mmol/L)	Confusion, dizziness Visual disturbances (e.g., diplopia) Slurred speech, loss of coordination Increased pain threshold Lethargy
250–400 mg/dL (54–87 mmol/L)	Inability to stand or walk Markedly decreased response to stimuli Vomiting, incontinence Stupor
350–500 mg/dL (76–109 mmol/L)	Depressed or absent reflexes Hypothermia Impaired respiration Unconsciousness, coma

Source: Information compiled from multiple sources.

are considered legally intoxicated. Alcohol concentrations exceeding 350 mg/dL (76 mmol/L) can be fatal.

Excessive long-term use of ethanol is associated with damage to a number of organ systems.

- Liver: fatty liver, alcoholic hepatitis, cirrhosis
- Gastrointestinal tract: acute and chronic pancreatitis, gastritis, malabsorption/malnutrition

- Nervous system: peripheral neuropathy, cerebral atrophy/dementia, Wernicke–Korsakoff syndrome
- Heart: cardiomyopathy, arrhythmias, hypertension
- Blood: anemia due to nutritional deficiency or bleeding
- Reproductive: fetal alcohol syndrome, testicular atrophy

Recreational and Therapeutic Uses

Most alcohol consumption is recreational. Beverages containing ethanol can be purchased by the bottle at liquor stores and by the drink at restaurants, bars, and pubs. The consumer decides how much he or she drinks, and responsible drinkers are expected to know their limits. Friends, relatives and bartenders sometimes intervene to cut off the supply of alcohol and prevent a bad outcome.

Alcohol is also used in several household products. The antimicrobial agent in many types of hand sanitizer gels is ethanol. The Centers for Disease Control and Prevention recommends using a hand rub with at least 60% ethanol content. Most brands of mouthwash contain ethanol as an antibacterial agent and as a solvent for other components.

One specific medical use of ethanol is as an antidote for methanol and ethylene glycol poisoning. All three alcohols are metabolized by the enzyme alcohol dehydrogenase, but ethanol has a higher affinity for the enzyme than the other two alcohols. Ethanol saturates the enzyme and prevents the metabolism of methanol and ethylene glycol to more toxic compounds.

Potential for Abuse

Alcohol is available to citizens of legal age and can be purchased at many locations in a community. By contrast, most drugs of abuse are either illegal to possess or are dispensed only by medical prescription. Alcohol is also part of mainstream culture – people are encouraged to drink alcohol in television commercials, in print media, and by their own peer groups. These facts help to explain why alcohol is the most frequently abused chemical substance among humans.

A standard drink is defined as 12 ounces of beer containing 5% alcohol, 5 ounces of wine containing 12% alcohol, and 1.5 ounces of distilled spirits containing 40% alcohol. The amount of alcohol in a standard drink is 14 g or 0.6 fluid ounces (18 mL).

The rate at which people become intoxicated depends upon their weight, gender, previous history of alcohol consumption and other factors. A large person requires more alcohol to reach a given blood concentration than a smaller person. Women become intoxicated more quickly than men because they have a greater proportion of body fat and a lower proportion of body water (alcohol distributes into body water compartments). People who drink alcohol every day metabolize ethanol more rapidly and develop tolerance to its intoxicating effects. The

(a)   **Approximate Blood Alcohol Content — Men**

BODY WEIGHT IN POUNDS

DRINKS	100	120	140	160	180	200	220	240
1	.04	.03	.03	.02	.02	.02	.02	.02
2	.08	.06	.05	.05	.04	.04	.03	.03
3	.11	.09	.08	.07	.06	.06	.05	.05
4	.15	.12	.11	.09	.08	.08	.07	.06
5	.19	.16	.13	.12	.11	.09	.09	.08
6	.23	.19	.16	.14	.13	.11	.10	.09
7	.26	.22	.19	.16	.15	.13	.12	.11
8	.30	.25	.21	.19	.17	.15	.14	.13
9	.34	.28	.24	.21	.19	.17	.15	.14
10	.38	.31	.27	.23	.21	.19	.17	.16

▨ Legally intoxicated

(b)   **Approximate Blood Alcohol Content — Women**

BODY WEIGHT IN POUNDS

DRINKS	100	120	140	160	180	200	220	240
1	.05	.04	.03	.03	.03	.02	.02	.02
2	.09	.08	.07	.06	.05	.05	.04	.04
3	.14	.11	.11	.09	.08	.07	.06	.06
4	.18	.15	.13	.11	.10	.09	.08	.08
5	.23	.19	.16	.14	.13	.11	.10	.09
6	.27	.23	.19	.17	.15	.14	.12	.11
7	.32	.27	.23	.20	.18	.16	.14	.13
8	.36	.30	.26	.23	.20	.18	.17	.15
9	.41	.34	.29	.26	.23	.20	.19	.17
10	.45	.38	.32	.28	.25	.23	.21	.19

▨ Legally intoxicated

Figure 13.2   Charts showing the effect of drinking on blood alcohol content (BAC) for (a) men and (b) women. The number of drinks consumed (left column) and the person's body weight (top row) determine the BAC. The numbers in each cell represent the BAC in g/dL 1 hour after the start of drinking. For every additional hour, subtract 0.015 g/dL from the number given to find the current BAC. Values are approximate due to individual differences in metabolism of ethanol.

relationship between alcohol intake and blood concentration is shown in separate charts for men and women in Figure 13.2.

Acute intoxication with alcohol impairs attention, judgment, coordination, and perception. People may become careless or aggressive. Many alcohol-related injuries are caused by falls, accidents, and violent behavior. A total of 10 511 deaths due to alcohol-impaired driving was recorded in the United States in 2018. When taken with other CNS depressants, such as opioids or benzodiazepines, alcohol can be a contributor to drug overdoses and deaths.

Long-term use of alcohol produces tolerance as well as physical dependence. Reduction or cessation of chronic alcohol use can trigger a withdrawal syndrome. Symptoms include tremors, sweating, nausea, vomiting, anxiety, insomnia, hallucinations, and seizures. Delirium tremens is the most severe form of alcohol withdrawal syndrome. In addition to the symptoms listed above, patients become confused, agitated, disoriented, feverish, and develop a rapid heart rate and high blood pressure due to instability of the autonomic nervous system.

In the 2018 National Survey on Drug Use and Health, 70% of respondents aged 18 years and older reported that they drank alcohol in the past year, and 55% reported consumption of alcohol in the past month. A total of 26.5% of people in this age group reported binge drinking in the past month, defined as five or more alcoholic drinks for males or four or more alcoholic drinks for females on the same occasion. A subset of this group (6.6% of respondents aged 18 years and older) engaged in heavy alcohol use, defined as binge drinking on five or more days in the past month.

Alcohol use disorder (AUD) is a recognized illness in which a person cannot control his or her alcohol use despite negative social, occupational or health consequences. The criteria for AUD, and the rating scale for mild, moderate and severe AUD are shown in Table 13.2. In 2018, 14.4 million adults aged 18 years and older in the United States (7.6% of men and 4.1% of women in this age group) had AUD. Another 400 000 adolescents aged 12–17 years (1.6% of this age group) had AUD. Only 8% and 5% of these groups, respectively, received treatment for AUD.

Metabolism

Following consumption, ethanol is absorbed in the duodenum and small intestine, then enters the bloodstream. Peak blood concentrations are reached in 30–60 minutes on an empty stomach. The peak occurs about 30 minutes later when the stomach is full, due to delayed gastric emptying. Alcohol is cleared from the blood at a rate of about 15 mg/dL (3.3 mmol/L) per hour, which is equivalent to the amount of alcohol in a standard drink.

More than 90% of ingested ethanol is oxidized by the liver. Alcohol dehydrogenase is an enzyme that converts ethanol to acetaldehyde, which is then further metabolized to acetate by aldehyde dehydrogenase (Figure 13.3). Ethanol is also metabolized to acetaldehyde by enzymes of the cytochrome P450 system. The latter pathway is induced by alcohol consumption and can increase the rate of ethanol metabolism in heavy drinkers.

Table 13.2 Criteria for alcohol use disorder according to the Diagnostic and Statistical Manual of Mental Disorders (DSM-5).

In the past year, have you:

- Had times when you ended up drinking more, or longer, than you intended?
- More than once wanted to cut down or stop drinking, or tried to, but couldn't?
- Spent a lot of time drinking? Or being sick or getting over other aftereffects?
- Wanted a drink so badly you couldn't think of anything else?
- Found that drinking – or being sick from drinking – often interfered with taking care of your home or family? Or caused job troubles? Or school problems?
- Continued to drink even though it was causing trouble with your family or friends?
- Given up or cut back on activities that were important or interesting to you, or gave you pleasure, in order to drink?
- More than once gotten into situations while or after drinking that increased your chances of getting hurt (such as driving, swimming, using machinery, walking in a dangerous area, or having unsafe sex)?
- Continued to drink even though it was making you feel depressed or anxious or adding to another health problem? Or after having had a memory blackout?
- Had to drink much more than you once did to get the effect you want? Or found that your usual number of drinks had much less effect than before?
- Found that when the effects of alcohol were wearing off, you had withdrawal symptoms, such as trouble sleeping, shakiness, restlessness, nausea, sweating, a racing heart, or a seizure? Or sensed things that were not there?

The presence of at least 2 of these symptoms indicates alcohol use disorder.

Severity is defined as: Mild: 2–3 symptoms; Moderate: 4–5 symptoms; Severe: 6 or more symptoms

---

Source: Reproduced from Alcohol use disorder: a comparison between DSM-IV and DSM-5. National Institute on Alcohol Abuse and Alcoholism. www.niaaa.nih.gov/publications/brochures-and-fact-sheets/alcohol-use-disorder-comparison-between-dsm.

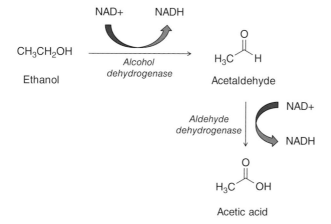

Figure 13.3 Metabolism of ethanol by the liver. Alcohol dehydrogenase converts ethanol to acetaldehyde, which is then metabolized to acetic acid. Both reactions generate equimolar amounts of NADH. Acetic acid is converted to acetyl CoA, which can enter the citric acid cycle to produce energy or contribute to the synthesis of fatty acids and cholesterol.

Table 13.3 Specimen type and window of detection for markers of alcohol consumption.

Marker	Specimen	Window of detection
Ethanol[a]	Serum/plasma/whole blood	6–8 h
	Urine	8–12 h
Ethyl glucuronide/ethyl sulfate	Urine	1–3 d
Phosphatidylethanol	Whole blood	2–3 wk
Carbohydrate-deficient transferrin	Serum	2–3 wk
Gamma-glutamyl transferase	Serum	2–3 wk
Aspartate aminotransferase/ alanine aminotransferase	Serum	2–4 wk
Mean corpuscular volume	Whole blood	2–4 mo

[a] Window of detection varies according to amount of ethanol consumed.
Source: Modified from Substance Abuse and Mental Health Services Administration. (2012). The Role of Biomarkers in the Treatment of Alcohol Use Disorders, 2012 Revision. *Advisory*, Volume 11, Issue 2. https://taadas.s3.amazonaws.com/files/36584446964799664-the-role-of-biomakers-in-the-treatment-of-alcohol-use-disorders.pdf

Unmetabolized ethanol is excreted in urine, breath, and sweat. A small proportion of ingested ethanol is converted to ethyl glucuronide and ethyl sulfate, which are excreted in urine.

Laboratory Testing

Tests for ethanol use fall into three categories: (i) direct measurement, (ii) analysis of metabolites, and (iii) indirect markers. The specimen to be analyzed can be serum/plasma, whole blood or urine, depending on which analyte is to be measured (Table 13.3).

Direct Measurement of Ethanol

Most clinical laboratories measure ethanol in serum or plasma with an automated chemistry analyzer. The method is based on converting ethanol to acetaldehyde by alcohol dehydrogenase and monitoring the change in absorbance at 340 nm as NADH is formed (Figure 13.3). Results are quantitative, reliable, and available within 1 hour.

Ethanol can also be measured by gas chromatography. This method is more time- and labor-intensive and is usually reserved for measuring ethanol in whole blood or as part of a toxic alcohol screen. The ratio of serum to whole blood ethanol is about 1.2, and this should be taken into account when interpreting test results.

Serum ethanol concentration is a useful guide to the degree of intoxication and is most frequently ordered in emergency departments. Ethanol concentrations in serum or whole blood are also measured for legal purposes (e.g., driving under the influence of alcohol). Values remain detectable for up to 6–8 hours following consumption. This figure may vary based on the amount of ethanol consumed and the rate at which it is metabolized by each individual.

Urine ethanol is measured in the same way as serum ethanol and provides evidence of consumption within the previous 8–12 hours.

Metabolites

A small fraction (<0.1%) of ethanol is conjugated to glucuronic acid and sulfate. The resulting compounds, ethyl glucuronide (EtG) and ethyl sulfate (EtS), are excreted in urine for 1–3 days following ethanol consumption (Figure 13.4). A screening immunoassay is available for EtG with a cut-off of either 500 or 1000 ng/mL. Positive results should be confirmed by LC-MS, which can also detect and measure EtS.

EtG and EtS are most useful for monitoring abstinence and identifying a relapse of drinking. When present at concentrations of 1000 ng/mL or greater, EtG indicates heavy drinking within the last 2 days or light drinking on the same day. The 500 ng/mL cut-off is more sensitive – a positive screening test at this level indicates heavy drinking within the last 3 days or light drinking in the past 24 hours.

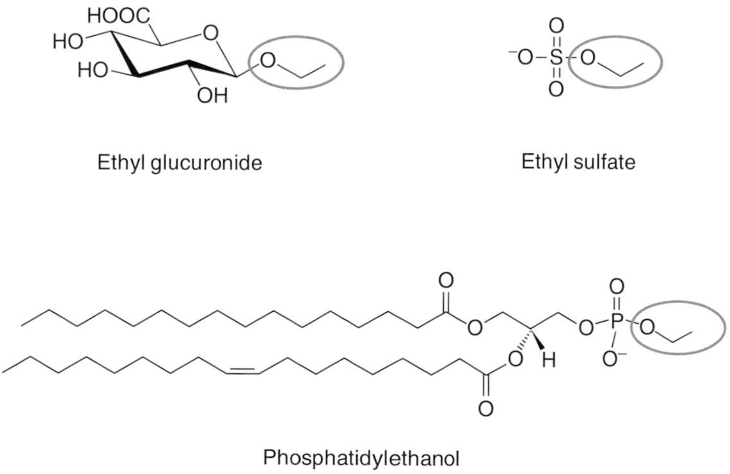

Ethyl glucuronide

Ethyl sulfate

Phosphatidylethanol

Figure 13.4 Structures of ethyl glucuronide, ethyl sulfate, and phosphatidylethanol. These metabolites are formed only in the presence of ethanol (circled), making them sensitive and specific markers of ethanol consumption. The fatty acid chains of phosphatidylethanol may vary in structure.

Recent use of alcohol-containing products such as mouthwash or hand sanitizer can occasionally produce a positive result in this lower range.

Other metabolites have been investigated as markers of alcohol intake. The most promising of these is phosphatidylethanol (PEth), a phospholipid formed only in the presence of ethanol (Figure 13.4). PEth values remain elevated for 2–3 weeks following alcohol consumption. The test is performed on whole blood and is not available at most clinical laboratories, so it is less convenient for routine monitoring than EtG.

Indirect Markers

The toxic effects of ethanol produce both qualitative and quantitative changes to plasma proteins, enzymes, and red blood cells. These indirect markers are useful in screening for alcoholism and may help to identify patients under treatment who relapse. The most widely used indirect markers are presented below.

Carbohydrate-deficient transferrin (CDT) is a fairly specific marker for recent alcohol consumption. Transferrin is the iron-binding protein that transports iron through the circulation and delivers it to tissues. It exists as a series of isoforms that differ from one another in the degree of glycosylation. The most common transferrin isoform contains two carbohydrate chains bearing four negatively charged sialic acid residues, but isoforms with 0–6 sialic acid residues may be present (Figure 13.5). The isoforms can be separated by capillary electrophoresis or high-performance liquid chromatography, and the relative amount of each isoform can be determined.

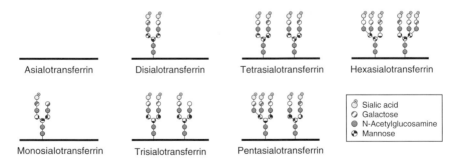

Figure 13.5 Diagram of transferrin isoforms. The protein backbone of transferrin is shown as a black line, to which one or two oligosaccharide chains are attached. Each chain consists of four different sugars, ending in a negatively charged sialic acid. The most abundant transferrin isoform in human serum is tetrasialotransferrin (containing four sialic acid residues). Consumption of alcohol increases the percentage of disialo- and asialo- isoforms. *Source:* Reproduced from Helander A, Wielders J, Anton R, et al. Standardisation and use of the alcohol biomarker carbohydrate-deficient transferrin (CDT). Clin Chim Acta 2016;459:19–24, with permission from Elsevier.

In patients who consume alcohol, there is an increase in the percentage of disialo- and asialo- transferrin isoforms (i.e., CDT). CDT is therefore used as an indicator of alcohol consumption. It is about as sensitive as GGT but is much more specific for alcohol intake. Liver disease, congenital disorders of glycosylation, and several genetic variants of transferrin can produce elevations in CDT as well. CDT is reported as a fraction of total transferrin – increased values (>2%) are indicative of alcohol use.

Gamma-glutamyl transferase (GGT) is a liver enzyme that is routinely measured by clinical laboratories. The activity of GGT in serum is elevated in most people who are chronic drinkers due to induction of the enzyme and release from damaged hepatocytes. It is a simple and fairly sensitive test to identify patients who relapse, as values rise quickly following alcohol consumption. Liver disease, biliary obstruction, and medications that induce GGT synthesis also cause an increase in activity, so the test is not a specific indicator of alcohol use.

Aspartate aminotransferase (AST) and *alanine aminotransferase* (ALT) are two more liver enzymes that are released in response to hepatic injury. Both are measured by automated chemistry analyzers and are available in nearly all clinical laboratories. Neither enzyme is specific for alcohol-mediated liver disease, so increased values are a reflection of hepatocellular injury from any cause. An AST/ALT ratio >2 suggests that the enzyme elevation is due to alcoholic liver disease.

Mean corpuscular volume (MCV), which gives the average size of red blood cells, is measured as part of a complete blood count. Individuals who consume at least several drinks of alcohol per day may show an increase in MCV. Other conditions that are associated with an elevated MCV include deficiencies of folate or vitamin B12, hypothyroidism, nonalcoholic liver disease, use of selected medications, and some hematological disorders. MCV is a less sensitive and specific test for alcohol use than CDT or GGT.

Because these markers perform differently, they are frequently ordered together to improve detection of alcohol use.

Window of Detection

Table 13.3 lists the window of detection for the most commonly used alcohol biomarkers.

Toxic Alcohols

Methanol, ethylene glycol, and isopropanol are short-chain alcohols that are present in a number of household products (Figure 13.1). These alcohols are occasionally ingested as a substitute for ethanol, by accident, or with suicidal intent.

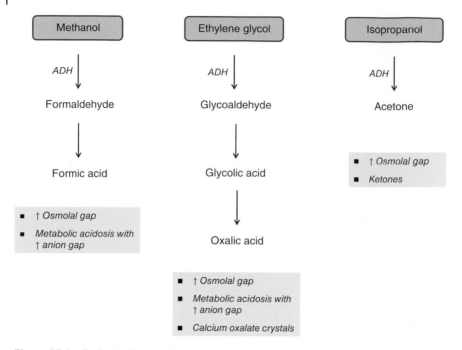

Figure 13.6 Toxic alcohols and their metabolites. For all three alcohols, the first step is catalyzed by alcohol dehydrogenase (ADH). Laboratory findings that accompany each type of intoxication appear in a shaded box below the pathway.

Like ethanol, they are rapidly absorbed in the stomach and small intestine, distribute into body water compartments, and cause inebriation and CNS depression.

These compounds are referred to as toxic alcohols because their consumption may produce serious or life-threatening consequences. Specifically, methanol and ethylene glycol are metabolized to toxic intermediates that cause acid–base and other biochemical abnormalities. Isopropanol induces sedation and coma at lower concentrations than ethanol. Figure 13.6 reviews the metabolism of toxic alcohols and their associated laboratory findings.

Methanol

Methanol is a component of many commercial solvents. Around the home, it is found in windshield washer fluid and small cans of cooking fuel that are used to heat food. Methanol is also added to some ethanol-containing products to prevent their consumption.

Following absorption, methanol is oxidized to formaldehyde by alcohol dehydrogenase. It is then converted to formic acid, which causes a metabolic acidosis and exerts a toxic effect on the optic nerve. Symptoms begin to appear 6–24 hours following ingestion, as formic acid accumulates. Patients develop abdominal pain, CNS depression, and blurry vision (sometimes compared to being in a snow-storm), which may progress to blindness.

Ethylene Glycol

Ethylene glycol is the main component in many types of automotive antifreeze and aircraft deicing solutions. It is a colorless, odorless liquid. Children and pets are attracted by the sweet taste, resulting in a number of accidental poisonings every year.

Ethylene glycol itself has limited toxicity and produces a level of inebriation similar to ethanol. Over a period of hours, it is metabolized to glycoaldehyde and then to glycolic acid, which produces a metabolic acidosis. Glycolic acid is further metabolized to glyoxylic acid and finally to oxalic acid, which forms an insoluble precipitate with calcium. Deposition of calcium oxalate crystals is responsible for tissue damage in multiple organs. Patients show neurological signs and symptoms within the first 12 hours, followed by cardiac and pulmonary dysfunction and acute kidney injury.

Isopropanol

Isopropanol is a widely used topical disinfectant, available as rubbing alcohol and in some brands of hand sanitizer. It is also used in cleaning solutions for eye-glasses, electronic equipment, and in the automotive industry.

Isopropanol is about twice as intoxicating as ethanol. It is metabolized by alcohol dehydrogenase to acetone, which also exerts a depressant effect on the CNS. Since acetone is not further metabolized, isopropanol ingestion does not produce a meta-bolic acidosis. Signs and symptoms are similar to ethanol intoxication – inebriation, decreased level of consciousness, and coma. Patients may suffer from nausea, vomiting, and abdominal pain due to irritation of the stomach by isopropanol.

Laboratory Testing

Direct Measurement of Toxic Alcohols

All the toxic alcohols, as well as ethanol, can be identified and measured by gas chro-matography. The lower limit of quantitation in serum or blood is usually 10 mg/dL, which is equivalent to 3.1 mmol/L methanol, 2.2 mmol/L ethanol, and 1.7 mmol/L isopropanol. Ethylene glycol is analyzed separately and can be measured at concen-trations as low as 20 mg/dL (3.2 mmol/L). The only metabolite that is routinely measured is acetone.

Gas chromatography requires special equipment and technical expertise, so it is not offered in many clinical laboratories. Even in full-service hospital labs, results may not be available for 2 hours or more. Fortunately, a number of routine tests are useful in identifying toxic alcohol ingestions.

Osmolal Gap

The presence of any alcohol in serum increases the osmolality, which is a measure of the number of solute particles in a solution. In serum, the main contributors to osmolality are sodium and its associated anions, glucose and urea. The osmolality of serum can be measured directly, and its expected level can be calculated by measuring the three analytes above and applying the following formulas:

$$2 \times \left[\text{sodium, mmol/L} \right] + \left[\text{glucose, mg/dL} \right] / 18$$
$$+ \left[\text{blood urea nitrogen, mg/dL} \right] / 2.8$$
$$= \text{calculated osmolality} \left(\text{mOsm/kg} \right)$$

or in SI units, with all values reported in mmol/L:

$$2 \times \left[\text{sodium} \right] + \left[\text{glucose} \right] + \left[\text{urea} \right] = \text{calculated osmolality} \left(\text{mmol/kg} \right)$$

The difference between measured and calculated osmolality (osmolal gap) is normally <10. An increased osmolal gap is evidence of an alcohol or another osmotically active substance in serum. All the above tests are available 24 hours/day in most clinical laboratories, and results are reported within an hour.

The osmolal gap is highest in the first few hours following an ingestion. Thereafter, metabolism and excretion of alcohols reduce the measured osmolality, with a corresponding drop in the osmolal gap.

Blood Gases, Anion Gap, Urinalysis

Metabolites of toxic alcohols can be detected with other common laboratory tests. Accumulation of formic acid (from methanol) or glycolic acid (from ethylene glycol) produces a metabolic acidosis. Blood gases reveal a pH <7.35, a low bicarbonate concentration, and a compensatory drop in pCO_2.

The anion gap, which is the difference between the major positively and negatively charged ions in serum, is calculated from electrolyte values:

$$\text{Anion gap} = \left[\text{sodium + potassium} \right] - \left[\text{chloride + total } CO_2 \right]$$

The anion gap is increased in methanol and ethylene glycol ingestions because their acidic metabolites consume bicarbonate (which is measured as total CO_2) and lower its concentration. Changes in blood pH and anion gap become more pronounced later in the course of an intoxication.

Additional clues come from analysis of urine. A positive test for ketones may indicate the presence of acetone, produced by oxidation of isopropanol. Calcium oxalate crystals on microscopic examination are consistent with ethylene glycol consumption. Figure 13.6 summarizes key laboratory findings in each type of ingestion.

Treatment

Both methanol and ethylene glycol poisoning can cause permanent damage or death if not recognized and treated promptly. The goals of treatment are to (i) block the production of toxic metabolites until the parent compound has been excreted, (ii) correct the metabolic acidosis, and (iii) provide supportive care throughout the acute episode. Ethanol competes for alcohol dehydrogenase and slows the rate at which either toxic alcohol is metabolized. Fomepizole is a specific inhibitor of alcohol dehydrogenase that is given for the same purpose. Therapy is continued until the concentration of methanol or ethylene glycol drops below a level considered safe. In severe cases, hemodialysis eliminates both the parent compound and its metabolites from blood.

For patients with isopropanol intoxication, supportive care is usually sufficient. Hemodialysis may be required in patients who are deeply comatose or have very high levels of isopropanol in their blood.

Case Studies

Case 13.1   Rough Weather

Captain Jonas Grumby was piloting his passenger ferry, the Squid, into port on the last run of the day. The seas were choppy and the wind gusted across the ship from port to starboard. As the ship neared the dock, it was tossed by a large wave into the outer set of pylons. The captain ordered full power to the engines, and the ferry crashed into the dock, damaging the loading ramp.

An investigation revealed that Captain Grumby had a blood alcohol concentration of 50 mg/dL (11 mmol/L) when the accident occurred. He admitted to keeping a flask of bourbon on the bridge to calm his nerves but claimed that bad weather and an unexpected surge of water were to blame.

- What are the regulations governing alcohol possession and use in the transportation industry?
- Is Captain Grumby in violation of the rules? Was he intoxicated?

(Continued)

Discussion

Employees in the transportation industry are subject to drug and alcohol testing if they perform safety-sensitive functions. These rules apply to many positions in the fields of aviation, railroads, and transit as well as to drivers of commercial vehicles and crew members operating a commercial maritime vessel. Regulations are established by the United States Department of Transportation.

Safety-sensitive employees must not use or possess alcohol while performing their jobs. They must not report for service or remain on duty if they:

- are under the influence of alcohol
- have a blood alcohol concentration of 40 mg/dL (9 mmol/L) or greater
- have used alcohol within 4 hours of reporting for service (8 hours for flight crews).

Based on these criteria, Captain Grumby broke the rules about possession and use of alcohol on the job, and his blood alcohol was higher than the allowable limit. He was mildly intoxicated, but less so than a legally intoxicated driver. Many experienced drinkers can function normally with this level of alcohol in their blood.

It may have been a case of bad luck. The weather was a factor in this accident – if the captain had successfully brought the ferry into the dock, no one would have checked his blood alcohol level. Safety-sensitive employees must submit to drug and alcohol testing if they are involved in an accident or crash that meets certain criteria. Employees may also be tested when they are hired, if there is reason to suspect that they are under the influence of drugs or alcohol, and on a random basis.

Takeaway messages

- Workplace testing for drug and alcohol use is standard practice in the transportation industry.
- US Department of Transportation regulations prohibit possession or use of illicit drugs and alcohol while on the job.
- Evidence of recent drug and/or alcohol use is sufficient cause for removing an employee from a safety-sensitive position.

Epilogue

After the hearing, Captain Grumby decided to leave his post with the ferry corporation, moved to Hawaii, and opened a charter sailing business for tourists. His new boat, the Minnow, disappeared during a storm with Grumby, his first mate, and five passengers aboard.

Case 13.2　A Clean Bill of Health?

WF, a 52-year-old lawyer, had an appointment with his doctor to discuss the results of some recent blood tests.

"Overall, you're in good shape," the doctor said. "Your blood glucose is normal, so diabetes is not a concern. Creatinine is also normal, and there's no protein in your urine, so your kidneys are doing fine. The lipid profile shows a borderline high cholesterol – not high enough for medication, but we'll keep an eye on it."

"That's encouraging," replied WF. "A clean bill of health?"

"Not quite. Your complete blood count had one blip, an increase in MCV. That means your red cells are larger than normal. And I checked an enzyme called gamma-GT. The value was high."

"Is that dangerous?"

"No, but it could indicate that you've been drinking more since I saw you last year."

- What questions should the doctor ask next?
- Are there any additional tests that should be ordered?

Discussion

Questions about alcohol consumption are a routine part of a patient's history. The CAGE* questionnaire is used in primary care to screen for alcohol abuse.

1) Have you ever felt you should <u>cut</u> down on your drinking?
2) Have people <u>annoyed</u> you by criticizing your drinking?
3) Have you ever felt bad or <u>guilty</u> about your drinking?
4) Have you ever had a drink first thing in the morning to steady your nerves or get rid of a hangover (<u>eye-opener</u>)?

If the patient answers yes to two or more questions, the screen is considered positive and may indicate excessive drinking or alcoholism.

A reliable history is worth more than a battery of lab tests in assessing a person's alcohol consumption. However, many people do not give an honest account of their drinking. In this case, there are two lab results suggesting that WF may be drinking more alcohol than is considered healthy.

Depending upon WF's responses to the CAGE questions, his doctor may decide not to order any additional lab tests now and repeat the mean corpuscular volume (MCV) and gamma-glutamyl transferase (GGT) in 3–6 months to see if they have normalized. Alternatively, he could order more tests to look for evidence of liver disease, which may be related to alcohol intake. A review of medications is also important, because certain drugs can induce GGT and increase its activity in serum.

(Continued)

The most useful test for long-term alcohol consumption is carbohydrate-deficient transferrin (CDT). Measurement of alcohol and its metabolite ethyl glucuronide is of limited value. These tests are more suitable for detecting alcohol consumption within hours to several days of collecting the specimen.

Takeaway messages

- Several commonly ordered lab tests may suggest a drinking problem.
- The best source of information about drinking is to ask the patient him/herself. If they answer honestly, the doctor can make appropriate decisions about ongoing care.
- There is no definitive laboratory test for alcoholism. CDT and GGT have good sensitivity and specificity for detecting long-term alcohol use.

*CAGE is an acronym derived from the first letters of the underlined words in each of the four questions.

Case 13.3 Preparing for Clinic

Nurse Boozer was reviewing lab results in preparation for that morning's alcohol use disorder clinic. Fifteen patients were scheduled – all but two had a negative screening test for ethyl glucuronide in their urine specimens. She looked at the results for the two patients who tested positive.

Analysis by LC-MS

Patient	Ethyl glucuronide (ng/mL)	Ethyl sulfate (ng/mL)
EB	185 000	34 000
JF	1600	250

A student nurse stopped at the desk, pointed to the result sheet and asked, "What are those lab tests for?"

Nurse Boozer responded, "We monitor all of our patients for alcohol use. It isn't easy to stop drinking, and many patients relapse at some point. We need to know how they are doing – otherwise they can slip back into old habits."

The student gave the nurse a quizzical look and said, "Oh … can you tell me what the results mean?"

- Why do laboratories test for ethyl glucuronide (EtG) and ethyl sulfate (EtS)?
- What does a positive test indicate?
- Are there any false-positive or false-negative results?

Discussion

EtG and EtS are metabolites of ethanol. Following consumption of ethanol, increased levels of both metabolites appear in urine for up to 3 days. Testing for EtG and EtS is an effective way to monitor patients in alcohol treatment programs.

Most screening immunoassays for EtG have a cut-off of 500 ng/mL and give either a positive or negative result. Confirmation testing is quantitative and measures both EtG and EtS. When positive, EtG and EtS are reliable indicators of drinking within the previous 1–3 days. The range of values for both metabolites can vary by more than a hundred-fold, as in the two cases above. These values are influenced by the amount of ethanol consumed and the time elapsed since the last drink.

Many brands of mouthwash and hand sanitizer have a high ethanol content. Heavy use of either product may cause the EtG screening test to become positive. Patients in alcohol treatment programs may be instructed to avoid ethanol-containing products, as they can interfere with test results.

Some people with recent alcohol intake will be missed by the 500 ng/mL screening cut-off. A lower cut-off of 100 or 200 ng/mL will detect a larger number of drinkers. However, this requires LC-MS analysis as the primary screening assay, and it increases the likelihood of a positive test from nonbeverage sources of alcohol.

In patients with urinary tract infections, glucuronidase produced by bacteria can degrade EtG in urine and give a false-negative result. EtS is not affected.

Takeaway messages

- Ethyl glucuronide is a good test for detecting alcohol use within the previous 1–3 days. Ethyl sulfate, which is present at lower concentrations, provides similar information.
- At the screening cut-off of 500 ng/mL, a positive EtG test is a reliable indicator of alcohol use. This cut-off is too high to pick up all recent drinkers.
- Use of an ethanol-containing mouthwash or hand sanitizer can produce occasional false-positive results.

Case 13.4 Trivia Night

Rob and Katie are regular customers of the Dew Drop Inn, a local pub that hosts Trivia Night every Thursday at 8 p.m. This week the couple arrived at 7.30. Rob ordered a pint of his favorite craft beer, and Katie requested a glass of red table wine. At 8 p.m. sharp, the quizmaster welcomed the crowd and

(Continued)

introduced the first topic, famous cartoon characters. This was followed by rock bands of the 70s, Italian Renaissance art, and great rivers of the world.

The evening was full of talk, laughter, more alcohol and a lot of wrong answers. Rob and Katie took their last sips just after 9.30 p.m. When the bill came an hour later, they tallied their consumption.

Rob – 3 pints of beer (16 oz), 1 shooter of tequila (1.5 oz)

Katie – 3 glasses of wine (5 oz)

- How many standard drinks did Rob and Katie each consume?
- Assuming that Rob weighs 180 pounds and Katie weighs 120 pounds, what would be the approximate blood alcohol content for each of them?
- Would they be considered legally intoxicated?

Discussion

The definition of a standard drink is 12 ounces of beer (5% alcohol), 5 ounces of wine (12% alcohol), and 1.5 ounces of distilled spirits (40% alcohol). Rob drank 48 oz of beer and 1.5 oz of tequila (distilled spirits) for a total of five standard drinks. Katie had 15 oz of wine for a total of three standard drinks.

The charts in Figure 13.2 give the approximate blood alcohol content an hour after consuming from one to 10 standard drinks. Three hours elapsed between their first sips of beer or wine and the time that they received the bill. Since alcohol is metabolized at a rate of 0.015 g/dL per hour, we need to subtract 0.015 g/dL × 2 extra hours, or 0.03 g/dL, to account for this.

The approximate blood alcohol content just before they left the pub was:

Rob: 0.11 g/dL (5 drinks for a 180-pound man) – 0.03 g/dL = 0.08 g/dL or 80 mg/dL

Katie: 0.11 g/dL (3 drinks for a 120-pound woman) – 0.03 g/dL = 0.08 g/dL or 80 mg/dL

According to these estimates, both Rob and Katie were legally intoxicated, because their blood alcohol content was at the threshold value of 80 mg/dL (17 mmol/L). Fortunately, the Dew Drop Inn is seven blocks from their house – they walked home.

Tables that estimate blood alcohol content are useful as a general guide, but they are only approximations. Not every pint of beer, glass of wine or jigger of spirits has the same alcohol content as a standard drink. The amount served may be more or less than the standard volume, and the percentage of alcohol may vary. In addition, people metabolize alcohol at different rates, which affects the postabsorption level in blood.

Takeaway messages

- You can estimate your blood alcohol content by keeping track of the number and type of drinks you consume. Calculators are available on the internet.
- Weight and gender are major determinants of blood alcohol content. In this case, Rob consumed 65% more alcohol than Katie, but their estimated blood alcohol values were the same.
- Don't drink and drive. Walk, take transit, call a taxi or appoint a nondrinking designated driver for the evening.

Case 13.5   Unconscious Man with an Osmolal Gap

An ambulance pulled up outside the emergency department of Big City General Hospital. When the rear doors opened, an unconscious middle-aged man in tattered clothing was wheeled into the building. The intern on duty asked the man his name, but there was no response.

"That's Freddy," the unit clerk said. "He's a frequent flyer here."

"Freddy," the intern repeated. "Hey, Freddy, wake up – you're at the hospital." Still no response.

The intern spoke to the paramedics, who said that Freddy was found passed out on a street corner in a rough part of town. He was well known to the ambulance service – this was his second ride to the hospital in the last 2 weeks.

"Did you find a bottle at the scene?" the intern asked.

"No," one of the paramedics said. "But I can tell you that he'll drink anything that smells like alcohol."

After quickly assessing her new patient, the intern collected one tube of venous blood for serum chemistry tests and a syringe of arterial blood for blood gas analysis. In under an hour, she had the following results.

Test	Result	Reference range
Sodium	139	135–145 mmol/L
Potassium	4.1	3.5–5.0 mmol/L
Chloride	105	98–108 mmol/L
Total CO_2	26	20–30 mmol/L
Glucose	4.7	3.3–11.0 mmol/L
Urea	6.2	2.0–9.0 mmol/L
Osmolality	325	270–295 mmol/kg
Ethanol	Not detected	Not detected
pH	7.38	7.35–7.45
pCO_2	42	35–45 mmHg
HCO_3^-	24	22–26 mmol/L

(Continued)

- What is the patient's calculated osmolality? Is the osmolal gap increased?
- What is the patient's anion gap? Is it elevated?
- Does the patient have a metabolic acidosis?
- Ingestion of which alcohol(s) could produce this pattern?
- What additional lab tests, if any, should the intern request?

Discussion

The calculated osmolality is the sum of: $[2 \times \text{sodium}] + [\text{glucose}] + [\text{urea}] = 289$ mmol/kg. If ethanol were present, it would be added to the above total. The osmolal gap is the difference between the measured and calculated osmolality:

$$325 (\text{measured}) - 289 (\text{calculated}) = 36 \, \text{mmol/kg}$$

This is well above the expected value of <10, so the osmolal gap is increased. (All serum chemistry results in the table above are reported in SI units to simplify the calculation of osmolality. In conventional units, glucose = 85 mg/dL and blood urea nitrogen = 17 mg/dL.)

The anion gap is the difference between the major cations and major anions in serum:

$$\left[139 + 4.1\right] - \left[105 + 26\right] = 12 \text{ mmol / L}$$

This value is normally 8–16 mmol/L, so the anion gap is not elevated. A slightly different formula omits potassium from the calculation, and the associated reference range is 4–12 mmol/L.

All of the blood gas parameters – pH, pCO_2 and HCO_3^- – are within the expected ranges. There is no disturbance of acid–base balance, so the patient does not have a metabolic acidosis.

The first set of calculations tells us that additional osmotically active substances are present in serum. Any of the three toxic alcohols could be responsible. The normal anion gap and blood gas results argue against methanol and ethylene glycol. Both compounds are metabolized to acidic intermediates that consume bicarbonate, lower the pH, and increase the anion gap. Isopropanol is therefore the most likely of the toxic alcohols to be present.

To confirm the diagnosis, alcohols by gas chromatography should be ordered. The serum sample drawn earlier was tested and showed the following.

Analysis by GC

Isopropanol	150 mg/dL	(25 mmol/L)
Acetone	52 mg/dL	(9 mmol/L)

Isopropanol is rapidly converted to acetone in vivo. The sum of their concentrations accounts for the increase in measured osmolality.

Takeaway messages

- Isopropanol is the most commonly ingested toxic alcohol. Rubbing alcohol and hand sanitizer are often the source of isopropanol.
- The key laboratory findings are an increased serum osmolality and osmolal gap with a normal pH and anion gap.
- Analysis by gas chromatography will detect isopropanol and acetone, but results may not be available until the next day. Patients can be managed without this information – repeat measurements of osmolality are a guide to the metabolism and excretion of isopropanol.

Case 13.6 Mass Poisoning in Estonia

During a 9-day period in September, 2001, 141 people presented to a community hospital in southwestern Estonia with signs and symptoms of methanol poisoning. Six more people were admitted to other hospitals. The most common findings on admission were gastrointestinal symptoms (nausea, vomiting, abdominal pain), visual disturbances, and difficulty breathing.

The community hospital's laboratory was not equipped to test for methanol. A presumptive diagnosis of methanol poisoning was based on history, clinical evaluation, and blood gas results that indicated the presence of a metabolic acidosis.

Fomepizole was not available in Estonia at that time, so patients were treated with ethanol to limit metabolism of methanol to formic acid. Sodium bicarbonate was administered to correct the metabolic acidosis, and mechanical ventilators were used to support respiration. Most patients were transferred to larger medical centers, where they received hemodialysis to remove methanol from their blood.

Despite these efforts, 25 patients died and 20 survived but had residual deficits, mainly related to impaired vision. Another 43 people died from methanol poisoning without ever getting to a hospital, bringing the total number of deaths to 68.

- How did this happen?
- Why do people drink moonshine?
- What is the relationship of blood methanol concentration and pH to patient outcome?
- Can anything be done to reduce the number of deaths?

(Continued)

Discussion

Outbreaks of methanol poisoning are usually due to consumption of illicit "bootleg" liquor, also referred to as moonshine. In this case, employees stole ten 200 L canisters of methanol from an industrial plant and sold it to an illicit alcohol trader. He mixed the methanol with water and lemon flavoring, bottled it, and attached fake labels for well-known brands of vodka. The bootleg alcohol was distributed through an underground network. Customers who drank the concoction became sick, and many of them died or suffered permanent disability.

Illicit alcohol can be produced cheaply and sold for less than liquor that is regulated by the government. This provides an economic incentive for people to drink moonshine. According to one estimate, bootleg alcohol accounted for more than one-third of the Estonian alcohol market. The lower price comes with less certainty about what the consumer is drinking.

A group of investigators collected clinical and laboratory data from the outbreak and published their findings several years later. Admission laboratory results for patients appear below, arranged according to clinical outcome.

Clinical outcome	Serum methanol[a]	Blood pH[a]
Survived	85 mg/dL (26.6 mmol/L)	7.19
Survived with residual deficits	229 mg/dL (71.4 mmol/L)	7.14
Died	352 mg/dL (110.0 mmol/L)	6.78

[a] Median values for serum methanol and blood pH. The range of values in each group was very wide. Methanol measurements were performed on admission blood samples retrospectively.
Source: Data from Paasma R, Hovda KE, Tikkerberi A, et al. Methanol mass poisoning in Estonia: outbreak in 154 patients. Clin Toxicol 2007;45:152–157.

These results demonstrate that higher serum methanol and lower blood pH values were associated with worse patient outcomes.

The key to surviving methanol poisoning is early recognition and treatment. Large overdoses can be successfully managed with fomepizole or ethanol and appropriate supportive care. Unfortunately, patients may equate symptoms of methanol poisoning with a hangover or other illness and choose not to seek help. Most of the deaths in Estonia occurred in people who did not receive medical attention.

Takeaway messages

- Mass poisonings with methanol have occurred in many countries through-out the world. Consumption of bootleg alcohol is the usual cause.
- Once the body starts to metabolize methanol, a metabolic acidosis develops. Blood gas measurements are a guide to the severity of the acidosis.
- Treatment decisions in suspected methanol poisoning cases need to be made quickly. Don't wait for serum methanol results – they may not be available for many hours.
- Untreated methanol poisoning carries a high risk of death and long-term sequelae.
- Despite the risks involved, moonshine remains a popular alcoholic beverage.

Further Reading

Articles

Andresen-Streichert, H., Müller, A., Glahn, A. et al. (2018). Alcohol biomarkers in clinical and forensic contexts. *Dtsch. Arztebl. Int.* 115: 309–315.

Kraut, J.A. and Mullins, M.E. (2018). Toxic alcohols. *N. Engl. J. Med.* 378: 270–280.

Report

Substance Abuse and Mental Health Services Administration. (2012). The Role of Biomarkers in the Treatment of Alcohol Use Disorders, 2012 Revision. *Advisory*, Volume 11, Issue 2. https://taadas.s3.amazonaws.com/files/36584446964799664-the-role-of-biomakers-in-the-treatment-of-alcohol-use-disorders.pdf

Book chapters

Wiener, S.W. (2015). Toxic alcohols. In: *Goldfrank's Toxicologic Emergencies*, 10e (eds. R.S. Hoffman, M.A. Howland, N.A. Lewin, et al.), 1346–1357. New York: McGraw-Hill.

Yip, L. (2015). Ethanol. In: *Goldfrank's Toxicologic Emergencies*, 10e (eds. R.S. Hoffman, M.A. Howland, N.A. Lewin, et al.), 1082–1093. New York: McGraw-Hill.

Websites

National Institute on Alcohol Abuse and Alcoholism
Alcohol Facts and Statistics
www.niaaa.nih.gov/publications/brochures-and-fact-sheets/alcohol-facts-and-statistics

Understanding the Dangers of Alcohol Overdose
www.niaaa.nih.gov/publications/brochures-and-fact-sheets/understanding-dangers-of-alcohol-overdose

National Highway Traffic Safety Administration

National Center for Statistics and Analysis. (2019) Alcohol-impaired driving: 2018 data. (Traffic Safety Facts. Report No. DOT HS 812 864) Washington, DC: National Highway Traffic Safety Administration.
https://crashstats.nhtsa.dot.gov/Api/Public/ViewPublication/812864

Centre for Addiction and Mental Health

Alcohol
www.camh.ca/en/health-info/mental-illness-and-addiction-index/alcohol
Alcohol and Other Drugs and Driving
www.camh.ca//en/health-info/guides-and-publications/alcohol-and-other-drugs-and-driving

Videos

2-Minute Neuroscience: Alcohol
www.youtube.com/watch?v=1D2uyrNcGuo
How Does Alcohol Make You Drunk?
https://ed.ted.com/lessons/how-does-alcohol-make-you-drunk-judy-grisel#watch

Section III

Appendices

Appendix A

How to Read a Toxicology Report

Test results for drugs are reported differently from other laboratory tests. Reports may appear lengthy or complex, because a single test shows up as multiple lines of information. Some of the terms – drug metabolites or analytical techniques – may be unfamiliar. This appendix gives an overview to reading and understanding a toxicology report.

Format

Reports can have a variety of formats, but the following elements are usually present.

At the top of the report, the patient's name, gender, and birthdate/age appear with numbers that identify both the patient and the specimen. The date and time of sample collection and name of the ordering physician are also given.

Reports may have a title that indicates the nature of the testing, such as Pain Management, Drugs of Abuse, Postmortem Toxicology, and so forth. This is followed by the test results and, in many cases, an interpretive comment.

The name, address, and contact information of the laboratory appear at the top or bottom of the report. A phone number for the laboratory may be listed here.

Screening Tests

Most screening assays are performed on urine samples. Individual tests are identified by the name of the drug or drug group, even though some assays measure a metabolite of the drug. Each test is reported as positive/negative or detected/not detected, and the cut-off value for a positive test is provided.

An Introduction to Testing for Drugs of Abuse, First Edition. William E. Schreiber.
© 2022 John Wiley & Sons Ltd. Published 2022 by John Wiley & Sons Ltd.

Screening assays are designed to give a positive result down to the cut-off concentration. A negative test usually means that the drug(s) of interest is not present. However, there are exceptions to this rule.

- Small amounts of a drug may be in the sample but do not reach the threshold for detection.
- When testing for a class of drugs, some compounds give a weaker signal than others. An example is oxycodone, which reacts weakly in the opiate assay and may give a negative result.
- Synthetic cannabinoids, opioids, and amphetamine derivatives are not detected with screening assays for these drug groups.

A positive screening test is usually reliable, especially when testing for a single drug such as cocaine or methadone. Consumption of certain drugs and foods can cause false positive results. See the articles by Moeller et al. (2017) and Saitman et al. (2014) in the reference section at the end of this appendix for a compilation of interfering substances.

When the results of a drug screen match the physician's expectations, there is usually no need for further testing.

Confirmation Tests

Confirmation of a positive screening result is required in several situations.

- The result is unexpected and has implications for patient care.
- The result may be used for legal purposes.
- Workplace drug testing programs.

Analyses are performed by gas chromatography-mass spectrometry or liquid chromatography-mass spectrometry techniques, which give detailed information on individual drugs and their metabolites. Samples are often pretreated with glucuronidase to remove glucuronic acid residues added by the liver. This step liberates the free drug or its metabolite and ensures maximum sensitivity of the assay.

A confirmation test report lists all drugs and metabolites that were detected and their concentrations. The cut-off for a positive test, which is typically lower than for screening tests, is also provided. Some laboratories include the analytical method as well.

Interpretation is more complex than for screening tests, because several compounds may be present in the sample. If a metabolite that is specific for a particular drug is detected, then the presence of that drug is confirmed. However,

Table A.1 Amphetamine-type stimulants, cocaine, and their metabolites.

Drug	Metabolites that indicate use
Amphetamine	
Methamphetamine	Amphetamine
Methylenedioxyethylamphetamine (MDEA)	Methylenedioxyamphetamine (MDA)
Methylenedioxymethamphetamine (MDMA)	Methylenedioxyamphetamine (MDA)
Methylphenidate	Ritalinic acid
Cocaine	Benzoylecgonine, cocaethylene[a]

[a] Cocaethylene is formed only in the presence of ethanol.

Table A.2 Benzodiazepines and their metabolites.

Drug	Metabolites that indicate use
Alprazolam	Alpha-hydroxyalprazolam
Chlordiazepoxide	Nordiazepam, oxazepam
Clonazepam	7-Aminoclonazepam
Clorazepate	Nordiazepam, oxazepam
Diazepam	Nordiazepam, temazepam, oxazepam
Flunitrazepam	7-Aminoflunitrazepam
Flurazepam	Hydroxyethylflurazepam, desalkylflurazepam
Lorazepam	
Oxazepam	
Temazepam	Oxazepam
Triazolam	Alpha-hydroxytriazolam

certain metabolites may be produced from more than one drug, and in some cases the metabolites themselves are drugs. When in doubt, call the laboratory and ask for an explanation of the results.

Drugs and Their Metabolites

Tables A.1–A.4 list the more common drugs of abuse by category and indicate which metabolites may be present following ingestion or administration of a particular drug. Not all parent drugs and their metabolites will be present in every case.

Table A.3 Opioids and their metabolites.

Drug	Metabolites that indicate use
Buprenorphine	Norbuprenorphine
Codeine	Morphine, hydrocodone
Fentanyl	Norfentanyl
Heroin	6-Acetylmorphine, morphine, codeine[a]
Hydrocodone	Norhydrocodone, hydromorphone, dihydrocodeine
Hydromorphone	
Meperidine	Normeperidine
Methadone	2-Ethylidene-1,5-dimethyl-3,3-diphenylpyrrolidine (EDDP)
Morphine	Hydromorphone
Oxycodone	Noroxycodone, oxymorphone, noroxymorphone
Oxymorphone	Noroxymorphone
Propoxyphene	Norpropoxyphene
Tramadol	O-desmethyltramadol

[a] When heroin is prepared from raw opium extracts, codeine may be detected in blood and urine samples. It is not a metabolite of heroin.

Table A.4 Alcohols, cannabis, and their metabolites.

Substance	Metabolites that indicate use
Ethanol	Ethyl glucuronide, ethyl sulfate, phosphatidylethanol
Isopropanol	Acetone
Tetrahydrocannabinol (THC)	11-nor-9-carboxy-THC (THC-COOH)

Further Reading

Moeller, K.E., Kissack, J.C., Atayee, R.S. et al. (2017). Clinical interpretation of urine drug tests: what clinicians need to know about urine drug screens. *Mayo Clin. Proc.* 92: 774–796.

Saitman, A., Park, H.-D., and Fitzgerald, R.L. (2014). False-positive interferences of common urine drug screen immunoassays: a review. *J. Anal. Toxicol.* 38: 387–396.

Appendix B

Guideline Documents: Pain Management and Addiction Medicine

Two fields in which drug testing plays an essential role are pain management and addiction medicine. Recent guideline documents that discuss drug testing in these areas of medical practice are summarized below.

Pain Management

A 2016 publication from the Centers for Disease Control and Prevention (CDC) presented 12 recommendations for prescribing opioids for chronic pain. Recommendation #10 relates to drug testing:

> When prescribing opioids for chronic pain, clinicians should use urine drug testing before starting opioid therapy and consider urine drug testing at least annually to assess for prescribed medications as well as other controlled prescription drugs and illicit drugs.

An expert panel convened by the American Academy of Pain Medicine used a consensus process to answer three questions related to urine drug monitoring (UDM) in patients receiving opioids for chronic pain. Their recommendations were published in 2018 and are summarized in Figure 2 of that paper. In abbreviated form, they are:

1) Perform definitive (i.e., confirmation) testing at baseline unless presumptive (i.e., screening) testing is required by institutional or payer policy.
2) Carry out a risk assessment for opioid misuse (patient history, validated tools, other sources of information).

3) On the basis of that assessment, perform UDM at least annually for low-risk patients, two or more times per year for moderate-risk patients, and three or more times per year for high-risk patients.

Both publications provide much greater detail than is presented here and are worth reviewing.

Addiction Medicine

The American Society of Addiction Medicine (ASAM) sponsored a consensus document that ". . . focuses on when, where and how often it is appropriate to perform drug testing in the identification, treatment, and recovery of patients with, or at risk for, addiction." The document was published in 2017 – it contains a detailed series of clinical recommendations to guide practitioners in the use of drug testing.

A pocket guide containing these recommendations is available on the ASAM website at: www.asam.org/Quality-Science/quality/drug-testing

Further Reading

Argoff, C.E., Alford, D.P., Fudin, J. et al. (2018). Rational urine drug monitoring in patients receiving opioids for chronic pain: consensus recommendations. *Pain Med.* 19: 97–117.

Dowell, D., Haegerich, T.M., and Chou, R. (2016). CDC guideline for prescribing opioids for chronic pain – United States, 2016. *MMWR Recomm. Rep.* 65 (No. RR-1): 1–49. www.cdc.gov/mmwr/volumes/65/rr/rr6501e1.htm.

Jarvis, M., Williams, J., Hurford, M. et al. (2017). Appropriate use of drug testing in clinical addiction medicine. *J. Addict. Med.* 11: 163–173.

Index

Page numbers in *italics* refer to figures; page numbers in **bold** refer to tables.

a

acetaldehyde   156, *157*
acetic acid   *157*
acetone   *162*, 163, 172–173
6-acetylmorphine   110
  case study   116–117
  cross-reactivity   **113**
addiction medicine   8, 183–184
adulterated specimen   23. *see also*
    adulteration under specimen
    validity testing
alanine aminotransferase (ALT)   161
alcohol. *see also* ethanol
  binge drinking   156
  bootleg   174–175
  content in alcoholic beverages   151
  effect of drinking on blood alcohol
    content (BAC)   *155*
  intoxication   **153**, 154–156
  markers of alcohol consumption
    158
  standard drink(s)   154, 170
  withdrawal syndrome   156
alcohol dehydrogenase   154, 156,
    157, 162
alcohol use disorder (AUD)   156, **157**

alprazolam (Xanax®)
  case study   90
  cross-reactivity   **82**
  dose, half-life, clinical use   **78**
  metabolite   *82*, **181**
  structure   *79*
American Academy of Pain
    Medicine   183
American Society of Addiction
    Medicine (ASAM)   184
amobarbital
  cross-reactivity   **96**
  half-life   **95**
  structure   *94*
amphetamine   62
  cross-reactivity   **67**
  structure   *60*
amphetamines
  case studies   69–74
  illicit drugs   63–65
  mechanism of action   59–60
  over-the-counter and prescription
    drugs   65–66
  physiological effects   60–61
  potential for abuse   61–62
  structure   59, *60*

An Introduction to Testing for Drugs of Abuse, First Edition. William E. Schreiber.
© 2022 John Wiley & Sons Ltd. Published 2022 by John Wiley & Sons Ltd.

amphetamines (*cont'd*)
   testing for   66–68
   therapeutic uses   61
   window of detection   67–68
anion gap   *162*, 164, 172–173
aspartate aminotransferase (AST)   161
attention deficit hyperactivity disorder
   (ADHD)   61, 72
ayahuasca   148

b
barbiturates
   case studies   102–104
   mechanism of action   93–94
   metabolism   95
   physiological effects   94
   potential for abuse   95
   structure   93, *94*
   testing for   96–97
   therapeutic uses   95
   window of detection   96
bath salts. *see* cathinones
benzocaine   54
benzodiazepines
   case studies   85–88, 89–91
   mechanism of action   77–78
   metabolism   81, *82*
   physiological effects   80
   potential for abuse   80–81
   structure   77, *78*, *79*
   testing for   81–83
   therapeutic uses   80
   window of detection   83
benzoylecgonine
   case studies   54–56
   cross-reactivity   **53**
   mass spectrum   *44*
   metabolite of cocaine   52
   screening tests   52
   structure   *50*
benzphetamine   68, 70

buprenorphine   112
   metabolite   **182**
   structure   *108*
butabarbital
   cross-reactivity   **96**
   half-life   **95**
butalbital
   cross-reactivity   **96**
   half-life   **95**
   structure   *94*
   therapeutic use   95
1,4-butanediol   *97*

c
cannabidiol (CBD)   127
   case study   132–133
   structure   *126*
cannabinoid receptor(s) (CB1, CB2)
   125, 127
cannabinoids
   case studies   130–135
   mechanism of action   125–127
   metabolism   128, *129*
   physiological effects   127
   potential for abuse   127–128
   source   125
   structure   125, *126*
   synthetic   128
   testing for   128–130
   therapeutic uses   127
   window of detection   130
cannabis
   case studies   130–135, 148–149
   consumption   128
   driving under the influence
     of   131–132
   medicinal   127
Cannabis sativa   125, *126*
carbohydrate-deficient transferrin
   (CDT)   160–161, 168
Catha edulis   64

cathinone *64*
cathinones 64–65
Centers for Disease Control and
    Prevention (CDC) 183
Centre for Addiction and Mental Health
    (CAMH) 11
chloral hydrate. *see also* trichloroethanol
    case studies 103
    metabolism 99
    structure *99*
chlordiazepoxide
    cross-reactivity **82**
    dose, half-life, clinical use **78**
    metabolites *82*, **181**
    structure *78*
chromatography 38. *see also* gas
    chromatography, liquid
    chromatography
    adsorption 38
    mobile phase 38
    stationary phase 38
Claviceps purpurea 137
clonazepam
    cross-reactivity **82**
    dose, half-life, clinical use **78**
    metabolite *82*, **181**
    structure *79*
cloned enzyme donor immunoassay
    (CEDIA) 29, *30*, 31
clorazepate
    dose, half-life, clinical use **78**
    metabolites *82*, **181**
cocaethylene
    case study 55–56
    formation 52
    structure *50*
cocaine
    case studies 53–58
    mechanism of action 49–51
    metabolism 52
    physiological effects 51

potential for abuse 51–52
source 49
structure 49, *50*
testing for 52–53
therapeutic uses 51
window of detection 53
cocaine metabolite. *see* benzoylecgonine
coca plant. *see Erythroxylon coca*
codeine 110
    case studies 116–117, 120–121
    cross-reactivity **113**
    metabolites **182**
    structure *106*
confirmation tests 37, 180
Controlled Drugs and
    Substances Act 6
crack cocaine 49, 51
cytochrome P450 (CYP) 15, 16

d
Department of Transportation, United
    States (US) 8, 166
    cut-off concentrations for drugs and
        metabolites in urine **9**
designer benzodiazepine(s) 90–91
dextroamphetamine
    (Dexedrine®) 59, 62, 72
dextromethorphan 148–149
diacetylmorphine. *see* heroin
Diagnostic and Statistical Manual of
    Mental Disorders (DSM-5) **157**
diazepam (Valium®)
    case study 87
    cross-reactivity **82**
    dose, half-life, clinical use **78**
    metabolites *82*, **181**
    structure *78*
dilute specimen 22, 23. *see also*
    dilution under specimen
    validity testing
N,N-dimethyltryptamine (DMT) 148

dissociative drugs
  case studies 145–147
  mechanism of action 143
  metabolism 144
  physiological effects 143–144
  potential for abuse 144
  structure 142, *143*
  testing for 144–145
  therapeutic uses 144
  window of detection 145
dopamine 49, *60*, 152
dronabinol 127
drug(s)
  controlled substance schedules 6, **7**
  definition 3
  economic impact of drug abuse 5–6
  harm associated with 4, *5*
  mechanism of action 3–4
  misuse and abuse 4
  reasons for testing 8–9
  terminology 4–5
Drug Enforcement Administration
  (DEA) 6, 127, 141
drug-facilitated sexual assault
  100–101, 144

e

ecgonine 49, **53**
ecgonine methyl ester *50*, 52
ecstasy. *see* methylenedioxymethamphetamine, MDMA
emergency medicine 8
endocannabinoids 125
enzyme multiplied immunoassay
  technique (EMIT) 29, *30*
Ephedra 65
Ephedrine 65, **67**
Erythroxylon coca 49, *50*
esketamine. *see* ketamine
eszopiclone. *see* zopiclone
ethanol. *see also* alcohol

case studies 165–171
laboratory testing 158–161
mechanism of action 152
metabolism 156–158
physiological effects 152–154
potential for abuse 154–156
recreational and therapeutic
  uses 154
source 151
structure *152*
window of detection **158**, 161
ethylene glycol 163
  metabolites *162*
  structure *152*
ethyl glucuronide (EtG) 159–160, 168–169
ethyl sulfate (EtS) 159, 168–169
etizolam 90–91
excited delirium 51, 57–58

f

fentanyl 111–112
  case study 115–116
  metabolite **182**
  structure *107*
flumazenil 81
flunitrazepam
  case study 100–101
  metabolite **181**
  structure *79*
flurazepam
  cross-reactivity **82**
  dose, half-life, clinical use **78**
  metabolites *82*, **181**
fomepizole 165
forensic toxicology 8

g

GABA-A receptor 77–78, 93–94, 152
GABA-B receptor 97
gamma-aminobutyric acid (GABA) 77–78,
  93–94, *97*, 152
gamma-butyrolactone *97*

gamma-glutamyl transferase (GGT)
161, 167–168
gamma-hydroxybutyric acid (GHB)
case studies   100–101
mechanism of action   97–98
metabolism   99
physiological effects   98
potential for abuse   98
source   97
structure   *97*
testing for   99
therapeutic uses   98
gas chromatograph   *39*
gas chromatography (GC)   39–41
carrier gas   39
chromatogram   39, *40*
column(s)   39, 40
detector(s)   39, 40–41
retention time(s)   39, *40*
gas chromatography-mass spectrometry
(GC-MS). *see* confirmation tests
GHB receptor   98
glucuronic acid   15
glutamate   93–94, 143, 152

h

hallucinogens   137
hallucinogens, classic
case studies   147–149
mechanism of action   140
metabolism   141
physiological effects   140–141
potential for abuse   141
source   137, 138
structure   138, *139*
testing for   141–142
therapeutic uses   141
window of detection   142
hemp   129, 132, 133
heroin, 110. *see also* 6-acetylmorphine
case study   116–117

cross-reactivity   **113**
metabolites   **182**
structure   *106*
Hofmann, Albert   137
hydrocodone   111
cross-reactivity   **113**
metabolites   **182**
structure   *106*
hydromorphone   111
case study   119
cross-reactivity   **113**
structure   *106*
11-hydroxy-THC (11-OH-THC)
cross-reactivity   **129**
plasma levels   *129*
structure   *126*

i

immunoassay(s)
antibody interference   32
competitive   27
cross-reactivity   32
enzyme   28–31
heterogeneous   28, *29*
homogeneous   28
lateral flow   *33*, 34
matrix effect   32
microparticle aggregation   31
noncompetitive   28
prozone   32
radioimmunoassay(s) (RIA)   28, *29*
isomer(s), d (dextrorotatory) and l
(levorotatory)   59
isopropanol   163
case study   172–173
metabolite   *162*, **182**
structure   *152*

k

ketamine   143
case study   146–147
half-life   **142**

ketamine (*cont'd*)
    metabolite   144
    structure   *143*
khat plant. *see Catha edulis*
kinetic interaction of microparticles in
       solution (KIMS)   31
K2   128

l

levamisole   55–56
liquid chromatography (LC)   41–43
    chromatogram   *42*
    detector(s)   42–43
    high-performance (HPLC)   41
    mobile phase   41, 42
    retention times   41, *42*
    stationary phase   42
liquid chromatography-mass
       spectrometry (LC-MS). *see*
       confirmation tests
lisdexamfetamine (Vyvanse®)   62, 72
Lophophora williamsii   138, *139*
lorazepam (Ativan®)
    case study   85–86
    cross-reactivity   **82**
    dose, half-life, clinical use   **78**
    structure   *79*
lysergic acid diethylamide (LSD)   137
    case study   147–149
    half-life   **142**
    metabolite   141
    structure   *138*

m

marijuana, *129*, 132. *see also* cannabis
marijuana metabolite. *see* 11-nor-9-
       carboxy-THC, THC-COOH
mass analyzer(s)   43, *44*, 45
    high resolution   45
    quadrupole   *44*, 45
    triple quadrupole   45
mass spectrometer   43, *44*
mass spectrometry (MS)   43–45
    mass spectrum   43, *44*
    mass-to-charge (m/z) ratio
       43, *44*
    parent ion   43, *44*
    selected ion monitoring (SIM)   45
mean corpuscular volume (MCV)
    161, 167
medical review officer (MRO)   69
meperidine   112
    metabolite   **182**
    structure   *107*
mescaline   138–140
    half-life   **142**
    structure   *139*
metabolic acidosis   *162*, 164, 172,
    173, 175
methadone   111
    case study   117–118
    metabolite   **182**
    structure   *107*
methamphetamine   62
    case study   69–70
    cross-reactivity   **67**
    metabolite   **181**
    structure   *60*
methanol   162–163
    case study   173–175
    metabolites   *162*
    structure   *152*
methcathinone   *64*, 65
methylenedioxyamphetamine (MDA)
    63, **67**
methylenedioxyethylamphetamine
    (MDEA)   63
    cross-reactivity   **67**
    metabolite   **181**
    structure   *63*

methylenedioxymethamphetamine
  (MDMA)   63
  case study   71, 148–149
  cross-reactivity   **67**
  metabolite   **181**
  structure   *63*
methylenedioxypyrovalerone (MDPV)
  64, 65, 71–72
methylphenidate   66
  case study   72
  metabolite   **181**
  structure   *65*
morphine   110
  case studies   116–117, 119, 120–121
  cross-reactivity   **113**
  metabolite   **182**
  structure   *106*
mushrooms. *see* psilocybin

n
nabilone   127
naloxone   112
  case study   114–116
  structure   *108*
National Institute on Drug Abuse
  (NIDA)   5, 11
25I-NBOMe   148
N-methyl-D-aspartate (NMDA) receptor
  143, 152
11-nor-9-carboxy-THC (THC-COOH)
  case studies   131, 134, 135
  cross-reactivity   **129**
  plasma levels   *129*
  structure   *126*
nordiazepam   *82*, 87
norepinephrine   49, *60*

o
opiates   105, 115–117, 121–122
opioid receptor(s)   105, 107

opioids
  agonist(s)   107
  antagonist(s)   107, 112
  case studies   114–122
  mechanism of action   105–107
  naturally occurring   110
  physiological effects   107–108
  potential for abuse   109
  relative potency   **108**
  semisynthetic   110–111
  source   105
  structure   105, *106*
  synthetic   111–112
  testing for   113–114
  therapeutic uses   108–109
  window of detection   114
opium   105, 117
osmolal gap   *162*, 164, 171–173
osmolality   164, 171–173
oxazepam
  case study   87
  cross-reactivity   **82**
  dose, half-life, clinical use   **78**
  metabolism   *82*
  structure   *79*
oxycodone   111
  case study   121–122
  cross-reactivity   **113**
  metabolites   **182**
  structure   *106*
oxymorphone   111
  cross-reactivity   **113**
  metabolite   **182**
  structure   *106*

p
pain management   8, 108, 183–184
Papaver somniferum   105, *106*
paramethoxyamphetamine (PMA)
  63–64, **67**

paramethoxymethamphetamine
(PMMA)   63–64, **67**
pentobarbital
case study   103–104
cross-reactivity   **96**
half-life   **95**
structure   *94*
therapeutic use   95
performance-enhancing drug(s)
8–9, 133
peyote   138, 140
pharmacokinetics   13
absorption   13–14
distribution   14
excretion   16
metabolism   15–16
phencyclidine (PCP)   142
false-positive immunoassay
result   **145**
half-life   **142**
structure   142, *143*
phenobarbital (Luminal®)
case study   102–103
cross-reactivity   **96**
half-life   **95**
structure   *94*
therapeutic use   95
phentermine   66
cross-reactivity   **67**
structure   *65*
phosphatidylethanol (PEth)   *159*, 160
phosphoadenosine phosphosulfate
(PAPS)   *15*
phytocannabinoids   125
point-of-care testing (POCT)   33–34
poppy seed(s)   113, 114, 116–117
propoxyphene   112
metabolite   **182**
structure   *107*

pseudoephedrine   65
case study   74
cross-reactivity   **67**
structure   *65*
psilocin   138
half-life   **142**
metabolite   141
structure   *138*
Psilocybe (cubensis)   138, *139*
psilocybin   138, *139*

r
receptors, 3. *see also* individual
receptor types
reversed-phase chromatography. *see*
liquid chromatography
reward system   4
route of administration   13–14

s
Salvia divinorum   148
screening tests   27, 179
secobarbital
cross-reactivity   **96**
half-life   **95**
structure   *94*
serotonin   137, *138*, 140
serotonin receptor(s)   140
specimens
blood   19
hair   21
meconium   21
oral fluid   20
postmortem   55–56
sweat   21
urine   20
specimen validity testing (SVT)   21–24
adulteration   22
assessing validity   22–23
dilution   22

interpretation of SVT   23–24

substitution   22

Spice   128

substance   4

Substance Abuse and Mental Health
Services Administration
(SAMHSA)   11, **23**, **24**, 34

substance use by physicians   146–147

substituted specimen, 22, 23. *see also*
substitution under specimen
validity testing

sulfate   15

t

tandem mass spectrometer (MS/MS)   45.
see also mass spectrometry

temazepam
case study   87
cross-reactivity   **82**
dose, half-life, clinical use   **78**
metabolite   *82*, **181**
structure   *79*

tetrahydrocannabinol (THC)   125
case studies   131–133, 135
metabolite   **182**
plasma levels   *129*
structure   *126*

thebaine   105

thin-layer chromatography (TLC)   38

thiopental
cross-reactivity   **96**
half-life   **95**
therapeutic use   95

toxic alcohols
case studies   171–175
laboratory testing   163–165
metabolism   *162*, 163
signs/symptoms   163
source   162, 163

structure   *152*, 161
toxic effects   162
treatment   165

toxicology report
confirmation tests   180–181
drugs and their metabolites   181–182
format   179
screening tests   179–180

tramadol   112
metabolite   **182**
structure   *107*

triazolam
cross-reactivity   **82**
dose, half-life, clinical use   **78**
metabolite   *82*, **181**
structure   *79*

trichloroethanol   99, 103

u

uridine diphosphate (UDP)
glucuronic acid   *15*

w

workplace drug testing   8

World Anti-Doping Agency
(WADA)   133

World Health Organization   4

z

zaleplon
dose, half-life   **85**
metabolite   85
structure   *84*

Z-drugs
case studies   88–89
mechanism of action   84
metabolism   85
potential for abuse   84
structure   83, *84*

Z-drugs (*cont'd*)
   testing for   85
   therapeutic uses   84
   window of detection   85
zolpidem
   dose, half-life   **85**
   metabolite   85

   structure   *84*
zopiclone
   case study   89
   dose, half-life   **85**
   metabolite   85
   structure   *84*